Advances in Experimental Medicine and Biology

Volume 1057

Advances in Microbiology, Infectious Diseases and Public Health

This book series focuses on current progress in the broad field of medical microbiology, and covers both basic and applied topics related to the study of microbes, their interactions with human and animals, and emerging issues relevant for public health. Original research and review articles present and discuss multidisciplinary findings and developments on various aspects of microbiology, infectious diseases, and their diagnosis, treatment and prevention.

The book series publishes review and original research contributions, short reports as well as guest edited thematic book volumes. All contributions will be published online first and collected in book volumes. There are no publication costs.

Advances in Microbiology, Infectious Diseases and Public Health is a subseries of *Advances in Experimental Medicine and Biology*, which has been publishing significant contributions in the field for over 30 years and is indexed in Medline, Scopus, EMBASE, BIOSIS, Biological Abstracts, CSA, Biological Sciences and Living Resources (ASFA-1), and Biological Sciences. 2016 Impact Factor: 1.881

More information about this series at http://www.springer.com/series/13513

Gianfranco Donelli

Editor

Advances in Microbiology, Infectious Diseases and Public Health

Volume 9

 Springer

Editor
Gianfranco Donelli
Microbial Biofilm Laboratory
Fondazione Santa Lucia IRCCS
Rome, Italy

ISSN 0065-2598   ISSN 2214-8019   (eBook)
Advances in Experimental Medicine and Biology
ISSN 2365-2675   ISSN 2365-2683   (eBook)
Advances in Microbiology, Infectious Diseases and Public Health
ISBN 978-3-030-07706-8   ISBN 978-3-319-79017-6 (eBook)
https://doi.org/10.1007/978-3-319-79017-6

This Springer imprint is published by the registered company Springer International Publishing AG part of Springer Nature.
The registered company address is: Gewerbestrasse 11, 6330 Cham, Switzerland

Contents

Adv Exp Med Biol - Advances in Microbiology, Infectious Diseases and Public Health (2018) 9: 1–27
DOI 10.1007/5584_2017_34
© Springer International Publishing AG 2017
Published online: 8 April 2017

Detection of Biofilms in Biopsies from Chronic Rhinosinusitis Patients: *In Vitro* Biofilm Forming Ability and Antimicrobial Susceptibility Testing in Biofilm Mode of Growth of Isolated Bacteria

Mariagrazia Di Luca, Elena Navari, Semih Esin, Melissa Menichini, Simona Barnini, Andrej Trampuz, Augusto Casani, and Giovanna Batoni

Abstract

Chronic rhinosinusitis (CRS) is the most common illness among chronic disorders that remains poorly understood from a pathogenic standpoint and has a significant impact on patient quality of life, as well as healthcare costs. Despite being widespread, little is known about the etiology of the CRS. Recent evidence, showing the presence of biofilms within the paranasal sinuses, suggests a role for biofilm in the pathogenesis. To elucidate the role of biofilm in the pathogenesis of CRS, we assessed the presence of biofilm at the infection site and the ability of the aerobic flora isolated from CRS patients to form biofilm *in vitro*. For selected bacterial strains the susceptibility profiles to antibiotics in biofilm condition was also evaluated.

Staphylococci represented the majority of the isolates obtained from the infection site, with *S. epidermidis* being the most frequently isolated species. Other isolates were represented by *Enterobacteriaceae* or by species present in the oral flora. Confocal laser scanning microscopy (CLSM) of the mucosal biopsies taken from patients with CRS revealed the presence of biofilm in the majority of the samples. Strains isolated from the specific infection site of the CRS patients were able to form biofilm *in vitro* at moderate

M. Di Luca (✉)
Department of Translational Research and New Technologies in Medicine and Surgery, Pisa, Italy

NEST, Istituto Nanoscienze-CNR and Scuola Normale Superiore, Pisa, Italy

Berlin-Brandenburg Center for Regenerative Therapies, Charité-Universitätsmedizin, Berlin, Germany
e-mail: diluca.mariagrazia@gmail.com

E. Navari and A. Casani
Department of Medical and Surgical Pathology, Otorhinolaryngology Unit, Pisa University Hospital, Pisa, Italy

S. Esin, M. Menichini, and G. Batoni
Department of Translational Research and New Technologies in Medicine and Surgery, Pisa, Italy

Microbiology Unit, Pisa University Hospital, Pisa, Italy

S. Barnini
Microbiology Unit, Pisa University Hospital, Pisa, Italy

A. Trampuz
Berlin-Brandenburg Center for Regenerative Therapies, Charité-Universitätsmedizin, Berlin, Germany

Center for Musculoskeletal Surgery, Septic Unit
Charité-Universitätsmedizin, Berlin, Germany

or high levels, when tested in optimized conditions. No biofilm was observed by CLSM in the biopsies from control patients, although the same biopsies were positive for staphylococci in microbiological culture analysis. Drug-susceptibility tests demonstrated that the susceptibility profile of planktonic bacteria differs from that of sessile bacteria in biofilms.

Keywords

Adherent bacteria · Antibiotic resistance · Sinus infection · Isothermal microcalorimetry · Confocal fluorescence microscopy

1   Introduction

Rhinosinusitis (RS) comprises a group of disorders characterized by inflammation of the mucosa of the nose and the paranasal sinuses (Lanza and Kennedy 1997). According to the duration of the symptoms, RSs are sub-divided into acute forms, in which symptoms last up to 12 weeks followed by complete resolution, and chronic forms in which symptoms persist beyond 12 weeks. Chronic rhinosinusitis (CRS), whose prevalence is estimated to be approximately 10% of the population in the Western world, is considered a multifactorial disease in which both host-related (e.g. anatomical features, immunological status, ciliary dysfunction, associated co-morbidities) and non-host-related factors (e.g. microbial agents, allergens) may play a role (Fokkens et al. 2012; Lam et al. 2015). The etiology and pathogenesis of CRS remain an area of active research. In this respect, it is still debated whether bacterial infections, in the form of biofilms, might be involved in the pathogenesis of CRS contributing to the inflammation, persistence, and recurrent exacerbations of the disease (Lam et al. 2015). Bacterial biofilms are communities of sessile bacteria embedded in a protective extracellular polymeric substance (EPS) they have produced, and exhibiting a phenotype markedly different from the

corresponding planktonic cells concerning the growth rate and gene expression (Donlan and Costerton 2002). It is becoming increasingly evident that biofilms are involved in a large proportion of infections of the human body and may form on host tissues/mucosa or on the surface of a variety of medical devices including central vascular or urinary catheters, endotracheal tubes, prosthetic cardiac valves, orthopedic or dental implants (Lebeaux et al. 2014). Hallmark of biofilm-associated infections is their dramatically reduced susceptibility to antimicrobial treatments that is considered a multifactorial process (Lebeaux et al. 2014). In addition to the classical mechanisms of resistance at the single cell level (e.g. activation of efflux pumps, production of enzymes that destroy antibiotics, mutations that alter antibiotic target sites), multiple biofilm-specific mechanisms operate simultaneously and make it difficult to completely kill cells in a biofilm, especially those situated in the deeper layers (Penesyan et al. 2015; Sun et al. 2013). Consequently, biofilm-associated infections usually do not resolve easily, requiring the mechanical or surgical removal of the biofilm-infected tissue or colonized implant.

In addition to the tolerance to antimicrobial treatments, biofilms also exhibit a high capacity to resist the clearance by host innate and adaptive immune responses (Lebeaux et al. 2014). Both of these factors play a major role in treatment failure and persistence of biofilm-associated infections.

Biofilm-associated infections are not only difficult to treat, but also to diagnose by routine, culture-based, microbiological techniques (Trampuz et al. 2007). Indeed, methods traditionally used in clinical microbiology laboratories are optimized for detecting and culturing infectious microorganisms growing in planktonic (free floating) forms and to test their susceptibility under planktonic growth conditions (Hoiby et al. 2015). Due to the marked ability of biofilm cells to adhere to the substrate and/or to each other via the EPS, biofilm-embedded microorganisms might be much more difficult to recover from clinical samples than planktonic ones. Hence, the culture of biofilm cells from infected tissues is often negative, despite the presence of large aggregates of bacteria surrounded by an extracellular matrix can be demonstrated on tissues or medical devices by microscopic techniques (Hoiby et al. 2015; Batoni et al. 2016). Failure to isolate microbial strain(s) possibly involved in the pathogenic process, in turn, further affects the outcome of the antimicrobial therapy, in that the clinicians lack a rational basis for antibiotic selection (Hall-Stoodley et al. 2012).

In order to gain further insights on the possible role of biofilm-infections in CRS and diagnostic procedures suitable to aid laboratory diagnosis of such infections (e.g. type and mode of sampling, sample processing procedures, culture conditions etc.), the purposes of the present study were: (i) evaluate the presence of bacterial aggregates resembling biofilm structures in biopsies from CRS patients selected according to stringent clinical criteria (Benninger et al. 2003; Rosenfeld et al. 2015); (ii) isolate and identify aerobic bacterial strains from biopsies and other clinical samples collected from diseased and non-diseased areas by microbiological diagnostic techniques; (iii) assess the *in vitro* biofilm-forming ability of bacteria isolated from infection sites in different experimental conditions; (iv) compare the antibiotic susceptibility profile of the isolated strains in planktonic form versus that obtained in biofilm state.

The consistent demonstration of structured sessile bacteria on biopsies from CRS patients, and their marked ability to form biofilms *in vitro*, when tested in optimized experimental conditions, corroborates the hypothesis that biofilms may play a role in the pathogenesis/chronicity of the disease. The results obtained also suggest that modifications of routine microbiological procedures might be necessary to increase the rate of culture positivity. Finally, the choice of an eventual antimicrobial therapy to treat/prevent CRS could benefit from evaluation of the antibiotic sensitivity profiles of sessile bacteria.

2 Materials and Methods

2.1 Study Group and Sample Collection

The study group consisted of 12 patients with CRS, undergoing functional endoscopic sinus surgery (FESS). In addition, 5 patients undergoing septoplasty surgery for a surgical correction of a deviated nasal septum and reduction of the volume of the inferior turbinate with no clinical or radiological evidence of sinus disease or allergic rhinitis were selected as negative controls (1C-5C, Table S1). CRS patients were recruited after an accurate medical examination, according to the guidelines of the Rhinosinusitis task force (Benninger et al. 2003; Rosenfeld et al. 2015). In particular, medical examination was aimed at verifying the inflammation status of the nose and of the paranasal sinuses and its duration (minimum 12 weeks), and the presence of at least two of the following symptoms: nasal blockage/obstruction/congestion, nasal discharge (anterior/posterior nasal drip), facial pain/pressure, reduction or loss of smell, endoscopic signs of the disease. Clinical diagnosis of CRS was confirmed by using computerized tomography (CT) allowing a differential diagnosis with other diseases with similar symptoms such as upper respiratory tract infections or allergic rhinitis (Desrosiers et al. 2011).

Exclusion criteria included pregnancy, immunodeficiency, and impairment of muco-ciliary function. Preoperative data collection included symptom scores, allergy status, paranasal sinus computed tomography (CT) scores, past medical history, smoking status, and nasal endoscopy findings. No patients had taken antibiotics, antifungals, or steroids in the 4 weeks prior to surgery.

Samples were collected at the Otorhinolaryngology Unit of the Department of Surgery and Medicine, at the Pisa University Hospital, Pisa, Italy.

Table S1 Demographic, clinical and collected sample data

Patients	Gender	Age	Diagnosis	Polyposis	Nasal swab	Specific Sinus swab	Sinus aspirate	Sinus Biopsy
#1	M	74	Left maxillary sinusitis	NO	ND	Left maxillary sinus mucosa	ND	Left Maxillary sinus mucosa
#2	F	40	Left spheno-ethmoidal sinusitis	YES	ND	Left spheno-ethmoidal sinus mucosa	Left spheno-ethmoidal sinus	Left spheno.-ethmoidal sinus, sinonasal polyps
#3	M	47	Right maxillo-ethmoidal sinusitis	NO	ND	ND	Maxillary sinus	Maxillo-ethmoidal sinus
#4	M	78	Sinonasal polyposis, bilateral maxillo-ethmoidal sinusitis	YES	ND	Sinonasal polyps, maxillo-ethmoidal sinus	ND	Sinonasal polyps, maxilla-ethmoidal sinus
#5	F	62	Sinonasal polyposis, bilateral maxillo-ethmoidal sinusitis	YES	ND	Maxillo-ethmoidal sinus	ND	Maxillary sinus, sinonasal polyps
#6	F	65	Left maxillary sinusitis	NO	YES (L)	Left maxillo-ethmoidal sinus	ND	Left maxillo-ethmoidal sinus
#7	F	75	Pansinusitis	NO	ND	Maxillo-ethmoidal sinus mucosa	Ethmoido-maxillary sinus	Maxillo-ethmoidal sinus mucosa
#8	M	31	Right maxillo-ethmoidal sinusitis	NO	YES (L/R)	YES	ND	Right maxillo-ethmoidal sinus
#9	F	28	Pansinusitis	NO	YES	Maxillo-ethmoidal sinus mucosa	ND	Maxillo-ethmoidal sinus
#10	M	41	Pansinusitis	NO	YES	Maxillo-ethmoidal sinus	ND	Maxillo-ethmoidal sinus
#11	F	40	Right maxillo-ethmoidal sinusitis	NO	YES (R)	Right maxillo-ethmoidal sinus	Right middle meatus	Right maxillo-ethmoidal sinus
#12	M	59	Left maxillo-ethmoidal sinusitis	NO	YES (L/R)	Left maxillo-ethmoidal sinus	ND	Left maxillo-ethmoidal sinus
#1C	F	41	Nasal septum deviation	NO	YES	ND	ND	Inferior nasal turbinate
#2C	F	40	Nasal septum deviation	NO	YES	ND	ND	Left maxillary sinus
#3C	M	25	Nasal septum deviation	NO	YES	ND	ND	Inferior nasal turbinate
#4C	F	55	Nasal septum deviation	NO	YES	ND	ND	Inferior nasal turbinate
#5C	M	22	Nasal septum deviation	NO	YES	ND	ND	Inferior nasal turbinate

ND: not-determined

Nasal mucosa biopsies were harvested from CRS and control patients at the time of FESS and nasal septoplasty, respectively, and were placed in a sterile container. One to five pieces (size: ≤ 5 × 5 mm) were taken from each patient for confocal laser scanning microscopy (CLSM) and microbiological examination. Only the biopsy of the patient #9, was analysed by microbiological examination and not by CLSM due to its small size. In addition, sinus swabs from the site of the surgery and/or sinus aspirates were collected for bacterial and fungal cultures/identifications. Extreme attention was paid to avoid contamination with normal flora during sample collection. For some patients a nasal swab was also collected to evaluate the nose normal flora.

The Human Ethics Committee of the Pisa University Hospital approved the study and all patients provided their informed consent to participate in the study.

2.2 Detection of Biofilm-Like Structures in Biopsies from CRS and Control Patients by CLSM

The presence of bacterial aggregates resembling bacterial biofilms on biopsy samples (processed as described in paragraph 2.3) was evaluated by CLSM analysis as previously described (Foreman et al. 2009, 2010; Psaltis et al. 2007). Briefly, fresh specimens were immersed in 1 ml of sterile MilliQ water (Millipore), within 3 h from collection and added with 1.5 μl of Syto9 (Thermo Fisher Scientific). After incubation in the dark at room temperature for 15 min, each biopsy was rinsed in sterile MilliQ water to remove the excess of stain and placed on glass bottom dishes (Willco Wells) for microscopy imaging. In the case it was not possible to perform the analysis immediately, biopsy samples were fixed with 4% PFA for 4 h at 4 °C, washed three times with PBS, and stored at 4 °C in PBS added with 0.05% (w/v) NaN_3 to prevent microbial contamination, until staining and imaging. Stained specimens were observed under TCS SP5 II (Leica) confocal microscope (interfaced with a 488 nm Argon laser), using a 63 × 1.25 NA water immersion objective. For each biopsy the entire tissue surface and depth were scanned during three independent observations for a total of 45 min. 10 μm Axial stacks in the Z plane, with a slice thickness of 1 μm, were taken through representative areas of biofilm. The analysis was not performed blind.

2.3 Isolation, Identification and Planktonic Drug Susceptibility Testing of Clinical Isolates

Biopsies, sinus swabs, and/or sinus aspirates were transferred within 1–3 h from collection to the Microbiology Unit laboratory of Pisa University Hospital, and processed as follows: biopsies were washed three times in PBS to remove planktonic bacteria and divided into two pieces with a scalpel under sterile conditions. One of the pieces was then processed for staining and imaging as described in paragraph 2.2. The other piece was homogenised in 1 ml of PBS using a bead beater instrument (Stomaker), and vigorously vortexed for 1 min at room temperature. When multiple biopsies from the same patient were collected, they were individually processed for CLMS imaging, while they were pooled and homogenized all together for microbiological analyses. To this aim, 0.1 ml aliquots of the homogenates were inoculated directly onto blood agar, chocolate agar, mannitol salt agar, MacConkey agar, and Sabouraud agar, and cultured for 24–48 h at 37 °C in aerobiosis. The remaining homogenised samples were inoculated in 5 ml of brain heart infusion broth (BHIB, Becton Dickinson). After 24 h incubation at 37 °C, turbid broth samples were sub-cultured onto the five agar plates listed above. Selected colonies were identified using the matrix-assisted laser desorption ionization-time of flight mass spectrometry (MALDI-TOF) system (Bruker) using procedures suggested by the manufacturer. Swabs from nasal mucosa or from diseased sites were streaked directly onto the agar plates and cultured as described above. Sinus aspirates were processed as biopsies omitting the homogenization step.

For all isolated bacterial strains routine antibiotic susceptibility tests were performed by semiautomatic systems (Vitek II, bioMérieux) and/or Sensititre (Thermo Scientific).

2.4 Cultivation and Storage of Bacterial Strains

A single colony of bacterial isolates and standard laboratory strains (*Staphylococcus epidermidis* ATCC35984 and *S. aureus* ATCC43300) used as controls for biofilm production, were inoculated into 5 ml of Tryptone Soy Broth (TSB) (Oxoid,). *Haemophilus parainfluenzae* and *H. paraphrophilus* strains were cultivated in *Haemophilus* Test Medium Broth (Becton Dickinson). *Actinomyces odontolyticus, Streptococcus spp., Corynebacterium spp.,* and *Moraxella spp.* strains were cultivated in BHIB. After overnight incubation at 37 °C in shaking conditions, an aliquot of the cultures was re-inoculated in the same medium and the cultures were allowed to grow to an OD_{600} of 1. Each culture was divided into 0.25 ml aliquots, labelled and stored at −80 °C until use.

2.5 Stock Dilutions and Storage of Antibiotics

The antibiotics vancomycin hydrochloride, amoxicillin, amikacin, levofloxacin, doxycycline (all purchased from Sigma-Aldrich) were diluted in MilliQ sterile water (Millipore) to obtain stock solutions of 100 mg/ml. Erythromycin (Sigma-Aldrich) was diluted in pure ethanol (50 mg/ml) while rifampicin and daptomycin were diluted in DMSO (100 mg/ml). All the stock solutions were divided in aliquots and kept frozen at −80 °C or at 4 °C until use, as suggested by the producers.

2.6 Microtiter Plate Biofilm Formation Assay and Crystal Violet (CV) Staining

In a first set of experiments, overnight cultures of selected clinical isolates were diluted to 0.05 OD_{600} in TSB or Mueller Hinton Broth 2 cation adjusted (MHCA) (Sigma-Aldrich) added with 1% (w/v) glucose (Sigma-Aldrich) (MHCA/Glu) or 1% (w/v) sucrose (Sigma-Aldrich) (MHCA/Suc).

For selected species/strains additional biofilm formation assays were performed in MHCA or MHCA/Gluc added with 1% (w/v) NaCl, or in MHCA/Glu added with 1% (v/v) pooled heat inactivated human plasma collected from healthy donors.

Two hundred µl of each diluted bacterial suspension was dispensed into flat-bottom polystyrene 96-well plates (Corning Costar). Wells with medium alone were set up as negative controls. Plates were incubated at 37 °C without shaking for 24 h. After incubation, wells were washed three times with PBS, air-dried for 15 min at room temperature, and CV (Sigma-Aldrich) was added to a final concentration of 0.5% (w/v). Following incubation for 15 min at room temperature, wells were rinsed with PBS until no blue was visible in the washing solution, air-dried and CV was extracted by incubation for 15 min with pure ethanol (Sigma-Aldrich) at room temperature. The OD was measured at a wavelength of 595 nm with the INFINITE F50 (TECAN) absorbance microplate reader. The assays were performed in triplicate, and the results expressed as mean OD_{595} ± the standard deviation (SD). Based on the values of OD_{595} obtained, all strains were classified for their ability to form biofilms into four categories (no, weak, moderate and strong producers) according to Stepanovic and co-workers (Stepanovic et al. 2007).

2.7 CLSM and Image Analysis of *In Vitro* Formed Biofilms

CLSM and image analysis were used to analyze the architecture and extracellular matrix components of biofilms formed *in vitro* by staphylococcal clinical isolates, as previously described (Brancatisano et al. 2014; Maisetta et al. 2016). To this aim, 300 µl of bacterial suspensions, diluted as described above for microtiter plate biofilm formation assay, was dispensed into an 8-well µ-Slide (Ibidi) and statically incubated at 37 °C for 24 h. Following incubation, medium was carefully removed from the wells, biofilms were washed three times with 300 µl MilliQ water, and stained with 300 µl Syto9 (1.5 nM) for observation with CLSM, as suggested by the producer (Thermo Fisher Scientific). For the characterization of the staphylococcal biofilm extracellular matrix, biofilms were incubated in the dark for 20 min with 300 µl of undiluted FilmTracer™ SYPRO® Ruby biofilm matrix (Life Technologies) containing 5.0 µg/ml Wheat Germ Agglutinin, Oregon Green® 488 Conjugate (WGA488) (Life Technologies). After staining, biofilms were washed three times with pure water and examined under a Leica CLSM, as described above. An argon laser was used to excite the fluorophores at wavelengths of 458 nm for SYPRO Ruby and 488 nm for WGA and syto-9. The following collection ranges were adopted: 500–540 nm (WGA488 and Syto9), and 605–650 (SYPRO Ruby). In a typical two-channel experiment, images were collected in sequential mode to eliminate emission crosstalk or bleed-through between the various dyes. Two independent experiments were performed and 10 images for each sample were collected for biofilm formation and matrix component analysis.

2.8 Colony Morphology on Congo Red Agar

Staphylococcal strains isolated from CRS patients were tested for polysaccharide intercellular adhesin (PIA) production by the Congo red agar (CRA) method as described by Freeman and co-workers (1989) with minor modifications. To this aim, bacterial strains were plated on CRA (Mueller–Hinton broth 21 g l − 1, 1% bacteriological agar (Oxoid), glucose 50 g l − 1, and Congo red 0.8 g l − 1 [Sigma-Aldrich]). Plates were incubated for 24 h at 37 °C and for an additional 24 h at room temperature. Following incubation, the color and the morphology of the colonies were macroscopically examined. Black/gray colonies with a dry crystalline morphology allowed the recognition of PIA-producing strains, while red/pink, and smooth colonies were considered to be PIA negative.

2.9 Colorimetric Antibiotic Susceptibility Testing of Bacterial Biofilms by Alamar Blue (AB)

Biofilm susceptibility testing was assessed by microplate AB assay as described by Pettit and co-workers (2005). Briefly, biofilms were let to form for 24 h in MHCA/Glu medium or, in the case of *S. aureus*, in MHCA/Glu added 1% with human inactivated plasma as described above. Biofilms were washed three times with PBS and 200 µl of two-fold serial dilutions of antibiotics in fresh MHCA/Glu medium were dispensed into each well. Biofilms, exposed to drugs for 24 h at 37 °C, were washed three times with PBS and 5 µl of the oxidation reduction indicator AB was added to each well. Then, plates were shaken gently and incubated for 1 h at 37 °C. Color changes in the wells, from blue to pink or purplish, were visually recorded at the end of the incubation to assess the minimum biofilm inhibiting concentrations (MBICs), defined as the lowest drug concentration resulting in ≤50% reduction of AB and a purplish well

60 min after the addition of AB (Pettit et al. 2005).

2.10 Antibiotic Susceptibility Testing of *S. aureus* and *P. mirabilis* Biofilms by Isothermal Microcalorimetry Assay

Biofilm antibiotic susceptibility testing was performed by isothermal microcalorimetry (IMC) according to the procedure described by Oliva and co-workers (2014). Briefly, bacterial biofilms were formed on porous glass beads having a diameter 2 to 4 mm, porosity 0.2 m^2/g and pore size 60 μm (Siran carrier; SiKUG 023/02/300/A; Schott Schleiffer AG, Muttenz, Switzerland). To this aim, beads were incubated for 24 h at 37 °C in MHCA with 3–5 colonies of *S. aureus* or 1 colony of *P. mirabilis* ($\approx$1x10^8 CFU). Then, beads were washed and incubated an additional 24 h at 37 °C in the presence of two-fold serially diluted antibiotics at concentrations up to 1024 μg/ml. Following a second wash to remove non-attached bacteria and antibiotics, each bead was transferred into a different ampoule containing fresh medium without any antibiotics. Finally, all the ampoules were air-tightly sealed and inserted into the microcalorimeter to quantify the heat flow produced by viable bacteria. An isothermal microcalorimeter (TAM III, TA Instruments, Newcastle, USA), equipped with 48 independent channels and a detection limit of heat production of 0.2 μW was used. Growth media with biofilm on beads but without antibiotic pre-treatment were used as positive controls, while growth medium with sterile beads served as negative control. Experiments were performed in triplicates for each strain and the heat flow was recorded for about 48 h.

The minimum biofilm eradicating concentration (MBEC) was defined as the lowest concentration that eradicated biofilm and led to an absence of re-growth after 48 h of incubation in the microcalorimeter (Oliva et al. 2014).

2.11 Data Analysis

OD_{595} data of CV experiments were collected by spectrophotometer, analysed using Magellan software (TECAN), and exported/plotted using Microsoft Excel program.

Digital images of the CLSM optical section were collected using LAS-AF software (Leica, Heidelberg, Germany) and processed by ImageJ software (NIH Imagej; http://rsbweb.nih.gov/ij/). Microcalorimetry data were analyzed using the manufacturer's software (TAM Assistant, TA Instruments, New Castle, DE). Figures were plotted using GraphPad Prism 6.01 (GraphPad Software, La Jolla, CA, USA).

2.12 Statistical Analyses

The statistical significance of the data was determined by Fisher's exact test. A two-tailed P value of <0.05 was considered statistically significant.

3 Results and Discussion

3.1 Study Group and Demographic Data

A total of 17 patients were recruited for this study. Among these, 12 were patients with CRS undergoing FESS, and 5 were negative controls (patients undergoing septoplasty surgery).

The CRS patients' group consisted of 6 females and 6 males, (median = 53, range 28–78 yrs), The wide age-range of the study group is in agreement with data from the literature reporting that although CRS affects persons of all age groups, its incidence increases with age, peaking at 50 years or greater (Peters et al. 2014). The control group was made of 3 females and 2 males (median = 40, range 22–55 yrs). From a clinical point of view, the CRS patients included 3 subjects with nasal polyposis and 9 patients without polyposis. The demographic data, the clinical diagnosis, and the types of samples collected for CLSM analysis and microbiological examination from each patient and control are reported in supplementary information (Table S1).

3.2 Detection of Bacterial Biofilms in Biopsies from CRS and Control Patients by Syto9 Staining and CLSM Analysis

Biopsies from sinus mucosa of CRS and control patients were analyzed for the presence of biofilms by CLSM following fluorescence staining. Analysis of the biopsies was performed using a previously described, easy, and rapid protocol for the search of bacterial aggregates in tissue samples with minor modifications (Foreman et al. 2009; Foreman et al. 2010; Psaltis et al. 2007). The used probe specifically stains DNA of living and dead cells, including epithelial cells and bacterial cells, as well as the biofilm exopolysaccharide matrix, which contains bacterial DNA. Discrimination of human cells from bacterial cells is based on the different size and shape of the stained cells and,

in some cases, to differences in fluorescence intensity between bacterial and host cells, since bacteria stain brighter as compared to eukaryotic cells.

Seven out of 11 CRS patients analyzed by CLSM had biofilms detectable in their biopsies. Figure 1 shows representative magnified images of biofilm-like structures detected in the biopsy samples of patient #1, #4, #5, #6 and #12. The characteristic biofilm morphology with brightly fluorescent microorganisms, whose size is compatible with that of bacterial cells (ranging from 0.5 to 3 μm in most of the cases or longer) is clearly visible. In 3 out of 7 CRS patients whose biopsies were found biofilm-positive, only cocci or filament shaped bacteria were detected. In contrast, CLSM images of biopsies from patients #4 (Fig. 1B), #5 (Fig. 1C–D) and #12 (Fig. 1F) revealed the presence of bacterial cells with two different shapes (cocci and filaments), consistent with the isolation from these biopsies of *Staphylococcus lugdunensis/Citrobacter koseri* (#4), *S. aureus/Proteus mirabilis* (#5), and *S. epidermidis, S. capitis/Haemophilus parainfluenzae, Corynebacterium propinquum* (#12), respectively (Table 1). None of the biopsies from the control group was found

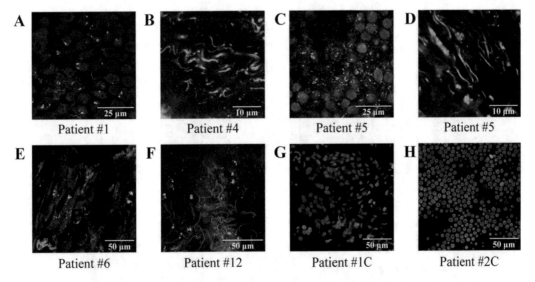

Fig. 1 CLSM imaging of sinus biopsies from patients #1, #4, #5, #6, #12 and controls #1C and #2C. The presence of bright green biofilm-like structures stained with Syto9 is evident in biopsies from CRS patients, but not from control patients. Aggregates of cocci shaped microorganisms (**A–C–E**). Aggregates of both cocci shaped microorganisms and long filamentous bacterial cells (**B–D–F**). Syto 9 staining (488/500–550 nm)

Table 1 Microorganisms isolated from CRS and control patients

Patients	Nasal swab	Sinus swab from diseased area	Sinus aspirate	Sinus Biopsy	Sinus Biopsy in BHI broth
#1	ND	*S. epidermidis*	ND	negative	*S. epidermidis*
#2	ND	*S. liquefaciens*	*S. liquefaciens*	negative	*S. liquefaciens*
#3	ND	ND	*S. aureus*	*S. aureus*	*S. aureus*
#4	ND	*S. lugdunensis, C. koseri*	ND	negative	*S. lugdunensis, C. koseri*
#5	ND	*P. mirabilis, S. aureus*	ND	negative	*P. mirabilis, S. aureus*
#6	negative	*S. anginosus, A. odontolyticus*	ND	negative	*S. epidermidis, S. anginosus*
#7	ND	*S. epidermidis, S. oralis, R. mucilaginosa*	*S. epidermidis*	*S. epidermidis*	*S. epidermidis*
#8	*S. epidermidis, S. haemolyticus*	*S. epidermidis, S. haemolityycus*	ND	*S. epidermidis, S. haemolyticus*	*S. epidermidis, S. haemolyticus*
#9	*S. epidermidis*	*S. epidermidis*	ND	*S. epidermidis*	*S. epidermidis*
#10	*E. coli,*	*E. coli,*	ND	*E. coli, M. catharralis, S. epidermidis, C. pseudodiphteriticum, S. oralis*	*E. coli*
#11	negative	*S. aureus, S. epidermidis, S. oralis*	*S. salivarius, S. mitis, H. paraphrophilus*	negative	*S. aureus*
#12	*S. epidermidis*	*H. parainfluenzae, C. propinquum, S. epidermidis,*	ND	*H. parainfluenzae, C. propinquum, S. epidermidis, S. capitis*	*H. parainfluenzae, C. propinquum, S. epidermidis, S. capitis*
#1C	*S. aureus*	ND	ND	*S. aureus*	*S. aureus*
#2C	Negative	ND	ND	*S. epidermidis*	*S. epidermidis*
#3C	*S. epidermidis*	ND	ND	negative	*S. epidermidis*
#4C	*S. epidermidis*	ND	ND	*S. epidermidis*	*S. epidermidis*
#5C	*Streptococcus spp*	ND	ND	*S. epidermidis, M. nonliquefaciens*	*S. epidermidis, M. nonliquefaciens*

ND: Not-done

positive for biofilm-like structures (0/5). Figure 1G–H shows two representative examples of biopsies from control group (#1C and #2C).

In few cases, multiple biopsy samples (2–5) were collected from CRS patients, in order to increase the sensitivity of the detection of the bacterial aggregates in the tissue, and individually evaluated by Syto9 staining and CLSM analysis. Of note, in one case in which it was possible to obtain 5 different biopsy samples, 3 of them were found positive for biofilm-like structures, while 2 of them were negative, suggesting that increasing the number of biopsies from a patient may increase the sensitivity of the technique.

A first analysis of the surface of the tissue sample from patient #4 resulted negative for the presence of bacterial aggregates. A successive more accurate analysis of the sample was performed by Z-stack imaging. Z-stack imaging revealed the presence of biofilm-like bacterial aggregates in the inner layers of sample #4 (Fig. 1B), indicating that biofilms may form not only on the mucosal surface, but also in the deeper layers of the tissue as also highlighted by others (Hall-Stoodley et al. 2012; Shields et al. 2013). For this reason, Z-stack imaging was performed also for all the other samples. Even with this accurate analysis no biofilm was detected in negative controls as well as in 4 of the CRS patients (#3, #7, #10, and #11) that were negative for the presence of biofilms on the surface of their biopsies (data not shown).

The bacterial aggregates detected deep in the tissue of patient #4 might also represent intramucosal microcolonies that have been also previously reported in CRS patients at a higher rate than in healthy controls (Wood et al. 2012; Kim et al. 2013a). Intramucosal bacteria were found to exhibit similar genotypic and phenotypic features as compared to surface bacteria isolated from the same patient and were hypothesized to reflect host mucosal immune dysfunction (Kim et al. 2013b).

Altogether, the results from CLSM analysis following fluorescent staining revealed the presence of bacterial aggregates resembling biofilms in the majority of the biopsies from CRS patients and in none of the control patients (P < 0.05).

With regard to the staining protocol, Syto9 staining is considered a reliable method for biofilm detection, with excellent inter-observer reliability (Foreman et al. 2009). Furthermore, such method has the advantage of allowing the imaging of fresh tissues, thus not altering the biofilm architecture, and of requiring a short time (around 20 min) for sample preparation, making it suitable for analysis in routine microbiology laboratory.

Another method frequently used to detect biofilms on tissue samples is the CLSM associated with fluorescent in situ hybridization (FISH/CLSM), which employs specific probes to target complementary sequences of the microbial 16 s RNA (Mallmann et al. 2010; Lubbert et al. 2016). Depending on the probe used, this method allows the visualization of all the bacteria present (universal EUB probe) or the identification of single species of microorganisms (species-specific designed probe) forming the biofilm (Wecke et al. 2000; Mallmann et al. 2010). Although more accurate, FISH/CLSM technique is based on a time-consuming protocol requiring over 3 h to be completed (Foreman et al. 2009). Compared to this method, Syto9/CLSM is not species specific, but it gives a more rapid answer regarding biofilm presence or absence. For this reason, Syto9/CLSM has been proposed as the technique of choice for investigating the generic presence of biofilm in CRS patients, an information that could be of value for clinicians for its demonstrated impact on the post-operative course of CRS patients following endoscopic sinus surgery (Foreman et al. 2009; Singhal et al. 2010).

3.3 Isolation and Identification of Bacterial Strains from CRS Patients and Control Group by Microbiological Procedures

Diagnostic microbiological procedures aimed at the isolation of aerobic bacterial strains and fungi were adopted in this study. Overall, 34 different bacterial strains were isolated and identified (Table 1) from the clinical samples (11 nasal swabs, 11 sinus swabs, 4 aspirates and 17 tissue specimens) taken from the patients and the controls (Table S1). Clinical isolates belonged to a total of 11 different genera and 21 distinct species of Gram-positive (n°13 species) and Gram-negative (n° 8 species) bacteria.

Among the isolated microorganisms, the most frequently Gram-positive bacteria found in CRS patients belonged to the *Staphylococcus* genus with the following species distribution: *S. aureus* (3 strains), *S. epidermidis* (10 strains), *S. lugdunensis* (1 strain), *S. haemolyticus* (1 strain) and *S. capitis* (1 strain). These results are in agreement with previous culture-based investigations reporting a prevalence of staphylococci among strains isolated from clinical samples obtained from CRS patients (Brook 2006; Shields et al. 2013).

Almost one third of the isolated bacterial species were Gram-negative (e.g. *Citrobacter koseri, Escherichia coli, Haemophilus parainfluenzae, H. paraphrophilus* (recently renamed *Aggregatibacter aphropilus), Moraxella catharralis, M. nonliquefaciens, Proteus mirabilis, Serratia liquefaciens*). In two/third of the samples from CRS patients, more than one bacterial species was isolated (Table 1). *S. liquefaciens, P. mirabilis,* and *E. coli,* isolated from patients #2, #5 and #10 respectively, are rarely recovered from middle meatus samples from healthy individuals and therefore their recovery from symptomatic CRS patients suggests that they could play a pathologic role. As previously suggested (Brook 2011), these organisms may have been selected out following administration of antimicrobial therapy in patients with CRS. In particular, *P. mirabilis* was isolated from a patient who had suffered recurrent rhinosinusitis and numerous previous sinus surgeries. This observation is in accordance with a retrospective study by Brook and Franzier (2001), who found a correlation between the presence of Gram-negative bacteria in maxillary sinus aspirates and the history of previous sinus surgery in CRS patients. *Serratia* genus is not frequently found in community-acquired rhinosinusitis and, when isolated, *S. marcescens* is the most represented

species isolated from nosocomial-acquired rhinosinusitis. In this study, *S. liquefaciens* was isolated from an immunocompetent patient (#2) who had no history of previous sinus surgery or long hospital stay. Extensive search in the literature revealed that *S. liquefaciens* is not a common Gram-negative bacteria found in CRS. Indeed, there are only three studies reporting the isolation of *S. liquefaciens* in rhinosinusitis (Snyman et al. 1988; Coffey et al. 2006; Richter and Gallagher 2016), and in one of them the bacterium was isolated from a case of acute sinusitis (Snyman et al. 1988).

Most of the bacteria isolated in this study were members of the oral or nasal normal flora. It is believed that the communication of the sinuses with the nasal cavity through the ostia could enable microorganisms that reside in the nasopharynx to spread into the sinus and, after closure of the ostia, these bacteria may be involved in the inflammation process (Brook 2011). No fungi were isolated from any of the specimens examined in this study.

The CRS study group included three subjects with nasal polyposis and nine patients without polyposis. Nasal polyps can impair paranasal sinus ventilation and drainage by blocking the ostiomeatal complex (Brook 2011). A recent study, based on the analysis of biopsies by fluorescent staining and CSLM, has shown that biofilms were more prevalent in CRS patients with nasal polyposis (33/34) compared to those without polyposis (22/27) and controls (14/25) (Danielsen et al. 2014). In our series, all the patients with polyposis (3/3) and 4/8 patients without polyposis presented biofilm-like aggregates in their biopsies. It has been reported that the microbial flora of the sinus of CRS patients with polyposis is not different from that found in patients with CRS without polyposis (Brook and Frazier 2001; Kim et al. 2014). In agreement with these studies, although biofilm seems to be more prevalent in CRS patients with polyposis, no evident correlation between the presence of polyposis and the severity of the disease was observed in our study group.

S. aureus, S. epidermidis and *M. nonliquefaciens* were isolated from nasal swabs and biopsies obtained from the control group subjects in agreement with the fact that such bacterial species may be part of the normal flora of nasal mucosa, although no evidence of biofilm-like structures was found in any of the subjects of our control group.

The question of whether normal bacterial flora in the sinuses exists is controversial (Brook 2011). Different studies have shown that the paranasal sinuses are sterile in healthy subjects with no history of sinus disease (Cook and Haber 1987; Sobin et al. 1992; Abou-Hamad et al. 2009). In contrast, other authors reported the presence of microorganisms in uninflamed sinus of healthy volunteers. In this regard, an early study in 1981 reported that *S. pyogenes*, *S. aureus*, *S. pneumoniae*, and *H. influenzae* can be commonly isolated from aspirates of patients without sign of CRS, who underwent corrective surgery for septal deviation (Brook 1981). Later, Jiang and colleagues endoscopically evaluated the bacterial flora of normal maxillary sinuses and isolated/cultured various microorganisms from half of the patients (Jiang et al. 1999). Recently, not only the presence of a normal flora, but also that of biofilms was demonstrated on the healthy mucosa of paranasal sinuses by scanning electron microscopy (Mladina et al. 2010). The discordance of the results among different studies might be due to: (i) the lack of standardization of the procedures used to collect samples from the sinus cavity; (ii) the failure to sterilize the area through which the endoscope is passed; (iii) the different anatomic areas from which the samples were collected (i.e. ethmoid bulla, maxillary antrum, or middle meatus); (iv) the kind of sample (swab, aspirate or mucosal tissue); (v) difference in the imaging or microbiological procedures adopted (e.g. transport time of the specimen, modalities of specimen processing and culturing). Further systematic studies with well-standardized protocols will be needed to clarify these aspects.

Regarding the procedures adopted in this study for the isolation of bacterial strains from surgical tissues, it is noteworthy that in 6/12 of the biopsies from CRS patients and in 1/5 of those from control subjects, the mere homogenization and vortexing of the tissue samples yielded negative culture results with no or very few colonies grown on agar plate (patients #1, #2, #4, #5, #6, #11, and #3C, Table 1). In contrast, isolation of bacterial strains was possible after incubating the same samples in an enrichment broth, suggesting that, as outlined by others, standard microbiological procedures might not be very efficient to detach bacteria embedded in the biofilm layers. Another hypothesis is that *in vivo* biofilm-forming bacteria are in a dormant/ametabolic/unculturable state and undergo re-activation only after inoculum in a rich liquid medium.

Different diagnostic procedures have been proposed to improve the isolation efficiency of bacteria from biofilms (Trampuz et al. 2007; Jost et al. 2014; Portillo et al. 2015). For instance, in the case a biofilm-associated joint prosthetic infections is suspected, sonication of orthopedic prostheses before the inoculum of the specimens into agar plates has demonstrated to promote the release of biofilm cells into planktonic form, increasing the number of positive cultures (Trampuz et al. 2007; Jost et al. 2014; Portillo et al. 2015). Very recently, Trampuz and coworkers have proved that the introduction of a sonication step and the inoculation of the sonication fluid into blood culture bottles further increases the isolation efficiency compared with the conventional sonication fluid or intraoperative tissue cultures (Portillo et al. 2015). It is likely that the introduction of a sonication step in the processing of biopsies taken from CRS patients for whom a biofilm infection is suspected could also improve the isolation efficiency directly from the tissue, avoiding the phase of the enrichment in broth. Indeed, although this latter step may facilitate the growth of bacteria embedded in the biofilm, in the case of mixed infections it may also alter the relative proportion of the bacterial species present in the sample, favoring those with faster growth kinetics (e.g. patient #10). On the other hand, the sonication step needs an accurate evaluation and standardization of the adopted parameters (e.g. duration and intensity of the treatment) to avoid loss of infectivity of the sample due to the sonication itself.

Routine antibiotic susceptibility testing was also performed on all the strains isolated from CRS patients and controls by semiautomatic systems. In most of the cases, the clinical isolates showed a wide range of antibiotic susceptibility (data not shown). The susceptibility profiles of *S. aureus* #5 and *P. mirabilis* #5 are reported in Table 3.

It is worth mentioning that a variety of molecular microbiological techniques has been developed in recent years that may help in the laboratory diagnosis of suspected biofilm infections with negative culture results (Costerton et al. 2011). Among them, particularly promising seems the combination of nucleic acid amplification procedures with electrospray ionization mass spectrometry (PCR/ESI-MS) (Ibis molecular method). Such technique, is based on the use of multiple set of primers designed to reveal known as well unknown bacteria. Detection and identification of fungi or viruses, or the presence in the specimen of bacterial genes that control the resistance to specific antibiotics (e.g. the MecA gene present in MRSA) is also possible. The amplicons produced by PCR are weighted by mass spectroscopy and their precise weight is used to calculate their base composition that is unique for each set of primers. Due to the high sensitivity and flexibility, the procedure has been proposed for the routine diagnosis of biofilm infections (Costerton et al. 2011).

3.4   Ability of Clinical Isolates to Form Biofilm *In Vitro*

A microtiter plate assay followed by CV staining (Brancatisano et al. 2014), was used to investigate *in vitro* the biofilm-forming ability of selected bacterial isolates that were classified into four categories, "no", "weak", "moderate", and "strong" biofilm producer, as proposed by Stepanovic et al. (2007). The capacity of a microorganism to form a biofilm on a given surface is highly influenced by several factors including the nature of the surface or the environmental conditions (e.g. nutrients, pH, temperature) (Martin-Rodriguez et al. 2014). In order to evaluate the influence of the medium on biofilm-forming ability *in vitro*, we first compared the biofilm-forming ability of a selected group of clinical isolates in MHCA/Glu to that of bacteria grown in TSB/Glu medium. MHCA is usually used to test antibiotic susceptibility, but the majority of the standard biofilm formation assays are performed in TSB. No significant difference in biofilm formation was observed in the presence of TSB/Glu and MHCA/Glu (data not shown). Therefore, MHCA was chosen as a medium for the successive experiments.

Secondly, we compared the ability of selected strains to form biofilm in the presence of 1% glucose *versus* 1% sucrose. Sucrose is a fermentable disaccharide and, similarly to glucose, it serves as a substrate for the synthesis of the biofilm extracellular polysaccharide as it has been described especially for oral bacteria (Paes Leme et al. 2006). As shown in Table 2, for most, but not all of the tested strains, no evident difference was observed in biofilm-forming ability in the presence of 1% glucose *versus* 1% sucrose. Among the Gram-negative bacteria tested, *S. liquefaciens* #2 and *P. mirabilis* #5 were strong biofilm producers, while *C. koseri* #4 was a moderate producer in both media. Among the Gram-positive bacteria, *S. anginosus* #6 was a moderate producer in MHCA/Glu, while it was classified as a weak producer in MHCA/Suc. Both in MHCA/Glu and MHCA/Suc the three *S. epidermidis* strains isolated from CRS patients were moderate (#6, #7) or strong (#1) biofilm producers. The *S. epidermidis* ATCC35984 strain, used as a control, was confirmed as a strong producer. The *S. epidermidis* strain isolated from the control patient (#2C) was a moderate biofilm producer in both media. *S. lugdunensis* #4, another coagulase-negative *Staphylococcus* strain, also showed a moderate ability to produce biofilm in both media. Both *S. aureus* strains isolated from CRS patients (#3, #5) showed weak ability to produce biofilm *in vitro*, while the *S. aureus* ATCC 43300 strain,

Table 2 Biofilm production of each isolated strain according to Stepanovic classification, in the presence of different growth medium (Stepanovic et al. 2007)

Strains	MHCA/ Gluc	MHCA/ Suc	MHCA/1% NaCl	MHCA/Gluc/ NaCl	MHCA/Glu/ Plasma
S. epidermidis ATCC35984	+++	+++	+++	++	++
S. epidermidis #1	+++	+++	+++	+++	+
S. epidermidis #6	++	++	ND	ND	ND
S. epidermidis #7	++	++	ND	ND	ND
S. epidermidis #2C	++	++	ND	ND	ND
S. aureus ATCC43300	++	++	ND	ND	+++
S. aureus #3	+	+	−	−	++
S. aureus #5	+	+	−	−	++
S. aureus #1C	+++	++	ND	ND	+++
S. lugdunensis #4	++	++	−	+	++
S. anginosus #6	++	+	ND	ND	++
C. koseri#4	++	++	−	−	+
S. liquefaciens #2	+++	+++	−	−	++
P. mirabilis #5	+++	+++	−	−	+++

Categories of biofilm producers: +++ Strong producers, ++ moderate producers, + weak producers, − no producer

formed biofilm at a moderate level. The *S. aureus* strain isolated from the control patient (#1C) was a strong producer in MHCA/Glu, but a moderate producer in MHCA/Suc. Several studies have documented that ability of staphylococci to produce biofilm *in vitro* is not always fully expressed and often requires a modification of the growth broth, in order to be fully manifested (Arciola et al. 2002, 2015; Stepanovic et al. 2007).

For instance, supplementation of TSB with NaCl, which is a known activator of *sigB* gene locus regulating specific genes in stationary phase and under different stress conditions, led to increased biofilm formation and PIA synthesis by *S. epidermidis* 1457 (Knobloch et al. 2001). Moreover, in two different studies the *rbf* gene of *S. aureus*, a transcriptional regulator factor, was found to be involved in the positive regulation of the multicellular aggregation step of biofilm formation in response to glucose and NaCl (Lim et al. 2004; Cue et al. 2009). For these reasons, the ability of two strains of *S. epidermidis* (ATCC35984 and #1) and two strains of *S. aureus* (#3 and #5) to produce biofilm in presence of Gluc and/or NaCl was also tested. In the presence of 1% NaCl in the medium without glucose, a reduction of more than 50% in biofilm

production was observed for the *S. epidermidis* ATCC35984 strain as compared to MHCA/Glu (data not shown), although the strain remained a strong biofilm producer (Table 2). A similar picture was obtained also for the *S. epidermidis* clinical isolate (#1), although the reduction was less evident (data not shown). In the presence of both glucose and NaCl, a reduction in biofilm production was visible for *S. epidermidis* strains ATCC35984. On the contrary in the same experimental conditions *S. epidermidis* #1 was still classified as strong biofilm producer. In the case of *S. aureus* (#3 and #5), adding of NaCl to the growth medium decreased the ability to form biofilm, both in the presence or absence of glucose.

It has been reported that *S. aureus* strains with poor ability to form biofilms *in vitro*, may express marked biofilm-forming capacity *in vivo* (Cardile et al. 2014). Factors potentially causing discrepancies between *in vivo* and *in vitro* conditions include the presence of host proteins, such as those present in human plasma (Aly and Levit 1987; Vaudaux et al. 1989; Wagner et al. 2011; Bjarnsholt et al. 2013). Body fluids may contain plasma with variable protein content according to the body district. For example, in nasal secretions, the plasma

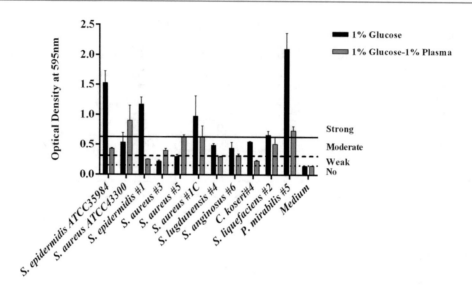

Fig. 2 *In vitro* **biofilm production by clinical isolates and ATCC strains of** *S. epidermidis* **and** *S. aureus* **in MHCA/Glucose or in MHCA/Glucose added with 1% of inactivated human plasma**. For each strain the biofilm biomass was evaluated after 24 h of incubation at 37 °C by CV staining and measurement of the optical density (OD) at 595 nm. Horizontal lines indicate the OD cut-off values used to define the categories of biofilm producers according to Stepanovic et al. (2007). OD values higher than the continuous line indicate strong biofilm producers

protein concentration ranges from 15 to 45% (Cardile et al. 2014) The importance of host proteins in facilitating biofilm formation is highlighted by studies demonstrating that medical implants are often coated by various host matrix proteins that enhance bacterial attachment to the surface and biofilm formation *in vivo* (Wagner et al. 2011; Bjarnsholt et al. 2013). *In vitro* the presence of plasma has been shown to promote biofilm growth as well (Chen et al. 2012; Walker and Horswill 2012). For this reason, the effect of 1% human plasma on biofilm formation was tested (Fig. 2, Table 2). Most of the tested *S. aureus* strains increased the production of biofilm in the presence of human plasma, passing from weak to moderate biofilm producers or, in the case of the *S. aureus* ATCC strain, from moderate to strong producer according to the categories of Stepanovic. Only the *S. aureus* #1C, isolated from one of the control patients, showed a slight reduction in biofilm formation but remained a strong biofilm producer (Fig. 2). A similar biofilm-inducing effect was not seen for the coagulase-negative staphylococcal strains (*S. epidermidis* ATCC35984 and #1, and *S. lugdunensis* #4) and for all the other strains tested (Fig. 2, Table 2). These data support the hypothesis that the supplementation of media with human plasma may better mimic the conditions encountered *in vivo* by certain, but not all the strains, favoring their ability to form biofilms *in vitro*.

The ability of a clinical isolate to form biofilm *in vitro* has been proposed as a predictive marker of its capacity to form biofilm also *in vivo* (Sanchez et al. 2013). Rapid and simple procedures for the routine analysis of biofilm forming ability of clinical isolates in the clinical microbiology laboratory have been recently proposed, with the aim to help the clinicians to identify high-risk infections and/or to predict the risk of therapeutic failure (Di Domenico et al. 2016). The results presented herein suggest caution in this respect, as they demonstrate that the ability to form biofilm *in vitro* might be highly influenced by the growing conditions adopted. Thus, poor ability to form biofilm *in vitro*, not necessarily is an indication of a low propensity of a strain to form biofilm also *in vivo*, but rather could indicate that the conditions adopted *in vitro* to grow that strain are sub-optimal.

3.5 Study of 3D Architecture and Extracellular Matrix Composition of Biofilms Formed *In Vitro* by Clinical Isolates

A more detailed characterization of biofilms formed *in vitro* by selected bacterial strains isolated from CRS patients was carried out after 24 h of growth in MHCA added with 1% glucose or, in the case of *S. aureus*, with 1% glucose and 1% human plasma. Figure 3A shows some representative CLSM images of biofilms stained with Syto9. The typical 3D architecture of biofilms, in which groups of microbial cells are separated by open water channels for the delivery of nutrients and oxygen, and the removal of metabolic waste is evident for most of the strains analyzed. In few cases, a peculiar pattern of bacterial distribution

or cell morphology was observed as in case of *S. liquefaciens*. In this case, characteristic filamentous cells, longer than their planktonic counterparts, were visible in agreement to what previously reported for the MG1 strain of *S. liquefaciens* (Givskov et al. 1998).

One of the main components of biofilms is the EPS, an abundant and complex matrix constituted by exopolysaccharides, proteins, extracellular DNA and other components that surround and protect biofilm cells (Donlan and Costerton 2002). The most characterized exopolysaccharide matrix component of *S. aureus* and *S. epidermidis* biofilms is PIA, also known as poly-N-acetyl-glucosamine (Arciola et al. 2015). When produced, PIA mediates adhesion to surfaces acting as a cementing matrix, enabling bacterial cells to

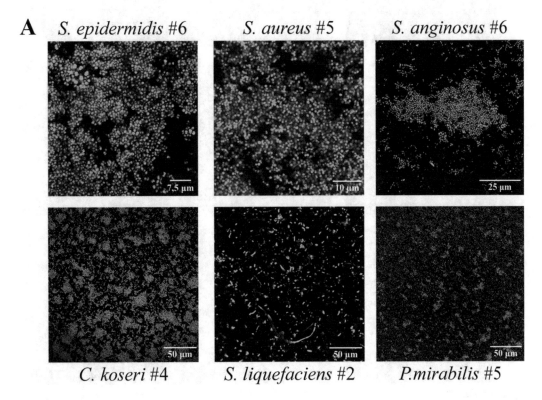

Fig. 3 *In vitro* **phenotypic characterisation of biofilms produced by different clinical strains.** (**A**) Architecture of biofilms produced *in vitro* by clinical isolates from CRS patients as visualized by Syto9 staining (488/500–550 nm) and CLSM analysis. Strains were cultivated in MHCA added with 1% glucose and, in the case of *S. aureus,* also with 1% human plasma. (**B**) Visualisation of biofilm matrix components of *S. aureus* and *S. epidermidis* clinical isolates by CLSM and double staining with WGA488 (detection of PIA in the green channel) and SpyroRuby (detection of proteins in the red channel). Scale bar: 10 μm

B

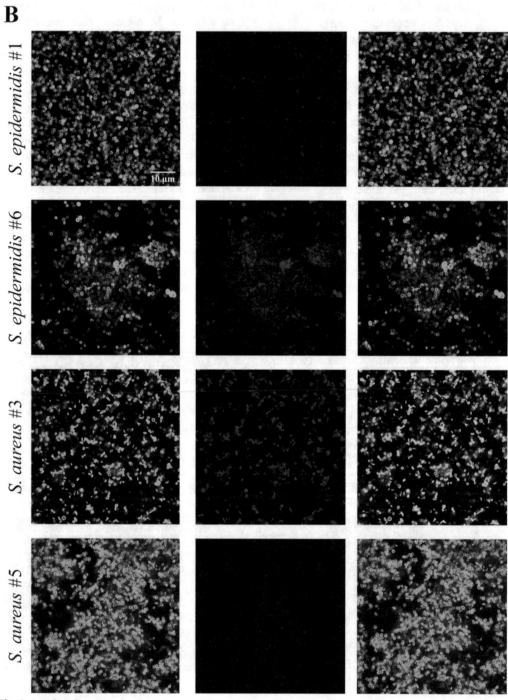

Fig. 3 (continued)

agglomerate in multi-layered biofilms, and making bacteria less accessible to the host defence system and antibiotics (Arciola et al. 2002). In order to phenotypically investigate the production of PIA by *Staphylococcus* strains isolated in our study, the colony morphology of such strains was analysed by the Congo Red Agar (CRA) plate method (Freeman et al. 1989). When grown on CRA, all the staphylococcal strains tested, but one, developed "black/gray" colonies indicative of a PIA-producing ability (see Fig. 4 for examples of colony morphology). The only staphylococcal strain that formed red colonies on CRA was *S. lugdunensis*, in agreement with the reported inability of this species to produce PIA (Arciola et al. 2015; Ravaioli et al. 2012). It has been reported that expression of PIA in *S. epidermidis* is subjected to phase variation due to the reversible insertion of an IS element in the operon encoding for PIA (Ziebuhr et al. 1999), although this mechanism has not been confirmed in later studies (Arciola et al. 2004). Phase variation of PIA expression has been described also for *S. aureus*. In *S. aureus* the phenomenon seems to be due to an expansion of the tetranucleotide tandem repeat housed in the PIA operon, resulting in a PIA-negative biofilm (Brooks and Jefferson 2014). It has been proposed that the ability to rapidly switch between phenotypes allows staphylococci to adapt to changing environmental conditions conferring an evolutionary advantage in the relationship with their hosts (Arciola et al. 2015).

In order to further characterise the biofilms of the staphylococcal strains isolated in this study in terms of PIA and/protein production, CLSM imaging of *S. aureus* and *S. epidermidis* biofilms was performed following double staining with the wheat germ agglutinin-Oregon green 488 conjugate (WGA$_{488}$) or with the red fluorescent dye Sypro Ruby to label PIA and proteins, respectively (Fig. 3B). In agreement with the results obtained by using the CRA method, all biofilms appeared green due to WGA488 binding to the N-acetyl-glucosamine residues of PIA in their matrix. Moreover, red fluorescence, due to Sypro Ruby binding, was also detectable in the

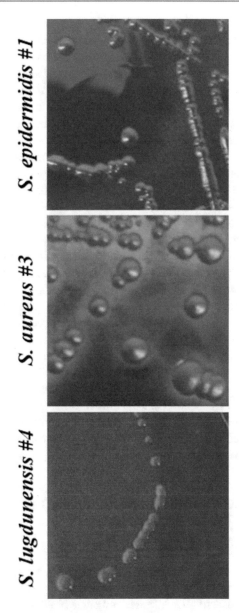

Fig. 4 Colony morphology on Congo red agar plates after 24 h incubation at 37 °C followed by 24 h incubation at room temperature. PIA producing ability was evaluated by visual detection of the colony color according to the chromatic scale of Arciola et al. (2002). Two PIA-positive strains (*S. epidermidis* #1 and *S. aureus* #3) and one PIA-negative strain (*S. lugdunensis* #4) are shown

biofilms of some strains (e.g. *S. epidermidis* #6, *S. aureus* #3, Fig. 3B), indicating the presence of protein components in the biofilm matrix of these strains. Interestingly, by using proteomic approaches, recently Gil and coworkers characterized the exoproteome of exopolysaccharide-based and protein-based biofilm matrices produced by two clinical isolates of *S. aureus* (Gil et al. 2014). They found that independently of the nature of the biofilm matrix, a common set of secreted proteins is contained in both types of exoproteomes.

A number of proteins localized in the extracellular matrix of biofilms have been described to be involved in generating PIA-independent biofilms. These include the biofilm-associated protein (Bap), a 2276-amino acid surface protein, and the accumulated associated protein (Aap) (Arciola et al. 2015). In addition, some authors suggest that various surface proteins called cell-wall anchored (CWA) proteins (including SasG, SasC, Protein A, FnBPA) can contribute to mediate the initial attachment and biofilm accumulation and maturation in *Staphylococcus* spp. (Merino et al. 2009; Vergara-Irigaray et al. 2009; Speziale et al. 2014). It has been proposed that staphylococci may produce PIA-dependent or PIA-independent biofilms in different phases of the pathogenesis of an infection and that this ability may allow the bacteria to better adapt to the multiple stimuli encountered in the host.

3.6 Antibiotic Susceptibility Testing of Bacterial Isolates in Biofilm Form

Several studies have indicated that activity of most antibiotics shows significant quantitative and qualitative differences against biofilm bacteria as compared to their planktonic counterparts. Therefore, it has been proposed that, when a biofilm-associated infection is suspected, the evaluation of antimicrobial susceptibility in biofilm form may better predict the success of the therapy and offers clinicians more appropriate guidelines to treat such infection than standard antimicrobial susceptibility assays (Macia et al. 2014; Hoiby et al. 2015). Although various methods have been described for testing and quantifying the activity of antibiotics against sessile bacteria over the last decade (Macia et al.

2014), to date, none of them have been approved as reference method in the clinical microbiology laboratory and there is a growing interest in the development of new susceptibility tests specific for biofilm-growing bacteria.

In a first set of experiments of this study, the colorimetric AB cell viability assay (Pettit et al. 2005) was used to test the antibiotic susceptibility of pre-formed biofilms of two Gram-positive (*S. aureus #5* and *S. epidermidis #1*), and two Gram-negative (*S. liquefaciens #2* and *P. mirabilis #5*) bacteria isolated from CRS-patients. Such assay is based on the evaluation of bacterial viability within the biofilm by the use of the blue dye resazurin. Reduction of the redox indicator by viable bacteria causes a colour change from non-fluorescent, blue (oxidized form) to fluorescent, red (reduced form) (Pettit et al. 2005; Tote et al. 2008). Two-fold serial dilutions of different antibiotics acting with distinct mode of actions were added to pre-formed (24 h–old) biofilms and incubated at 37 °C for 24 h. Then, AB was added to the wells and, after 1 h of incubation at 37 °C the MBIC was determined visually. MBIC was defined as the lowest drug concentration resulting in at least 50% reduction of AB that corresponded visually to a purplish coloured well 60 min after the addition of AB (Pettit et al. 2005).

As shown in Table 3 for *S. aureus #5* and *P. mirabilis #5*, overall MBIC values against biofilms of all tested strains were considerably higher than those obtained against planktonic bacteria by routine semi-automated antibiotic susceptibility testing. Of note, most of the antibiotics tested were not able to inhibit the metabolic activity of biofilms even at concentrations as high as 1024 µg/ml (Table 3).

Biofilm recalcitrance to antimicrobial treatment is a multifactorial phenomenon (Lebeaux et al. 2014) where inability of certain antibiotics to diffuse through the biofilm extracellular matrix may play a significant role. According to the charge and other physical-chemical features of both the antibiotic and the extracellular matrix, the antibiotic might be repulsed or alternatively sequestered by matrix components

Table 3 Antibiotic susceptibility of *S. aureus #5 and P. mirabilis #5* in planktonic form by Sensititre and VitekII systems and in sessile form by Alamar Blue assay (ABa) and isothermal microcalorimetry (IMC)

	S. aureus #5				P. mirabilis #5			
	MIC		MBIC	MBEC	MIC		MBIC	MBEC
	Sensititre	VitekII	ABa	IMC	Sensititre	VitekII	ABa	IMC
AMK	ND	ND	>1024	ND	<=4	4	256	256
AMP	<=0.25	R	>1024	ND	ND	ND	ND	ND
VAN	2	1	>1024	>1024	ND	ND	ND	ND
ERY	<=1	<=0.25	>1024	>1024	ND	ND	ND	ND
AMX	ND	ND	1024	ND	ND	ND	>1024	>1024
LEVO	0.25	ND	1024	1024	<=1	ND	<=4	512
DOX	<=0.05	ND	64	64	ND	4 (R)	128	512
RIF	<=0.06	<=0.025	<=2	128	ND	ND	ND	ND
DAP	0.5	ND	ND	256	ND	ND	ND	ND

ND: Not done
R: resistant
AMK: Amikacin, *AMP*: Ampicillin, *VAN*: Vancomycin, *ERY*: Erythromycin, *AMX*: Amoxicillin
LEVO: Levofloxacin, *DOX*: Doxycycline, *RIF*: Rifampicin, *DAP*: Daptomycin
MIC: minimum inhibiting concentrations
MBIC: minimum biofilm inhibiting concentrations
MBEC: minimum biofilm eradicating concentrations

hampering its penetration into the deeper biofilm layers (Lewis 2008; Lebeaux et al. 2014; Macia et al. 2014). Rifampicin was the most active antibiotic among those tested against staphylococcal biofilms, in agreement with the previously reported ability of such antibiotic to penetrate through the biofilm extracellular matrix (Dunne et al. 1993; Zheng and Stewart 2002). However, due to the rapid development of the resistance against rifampicin by staphylococcal strains, its use against biofilms is not recommended as mono-therapy, but in combination with other antimicrobials including tigecycline (Trampuz and Zimmerli 2006; Aslam and Darouiche 2007), linezolid (Raad et al. 1998; Saginur et al. 2006) and ciprofloxacin (Widmer et al. 1992; Zimmerli et al. 1998). The efficacy of the rifampicin against staphylococcal biofilms may suggest that targeting RNA polymerase is a highly effective strategy against the metabolic diversity of cells found in biofilms (Fey 2010).

Overall, a correlation was observed between the ability of an antibiotic to diffuse through the extracellular matrix and its ability to treat mature biofilms. For instance, although the strain of *S. aureus* tested (#5) was sensitive to vancomycin in planktonic form, biofilms of the same strain were highly resistant to vancomycin

(Table 3) in agreement with the low capacity of this antibiotic to penetrate *S. aureus* biofilms (Jefferson et al. 2005). Fluoroquinolones are reported to penetrate well into the extracellular matrix of biofilms formed by Gram negative species such as *P. aeruginosa* and *E. coli* (Lebeaux et al. 2014). In accordance, levofloxacin exhibited an antibiofilm activity at low concentrations (MBIC ≤ 4 μg/ml) against *P. mirabilis* tested in this study (Table 3). Doxycycline was able to inhibit biofilm metabolic activity of the same strain at a relatively low concentration (MBIC = 128 μg/ml), even if higher than that achievable in clinical practice.

IMC is a highly sensitive technique, which allows measuring the heat flow generated or consumed during a biological process such as bacterial growth (Braissant et al. 2010). IMC has been previously used with success for clinical and diagnostic purposes to test the activity of antibiotics and antimycotics against bacterial and fungal biofilms (Furustrand Tafin et al. 2012; Furustrand Tafin et al. 2013; Oliva et al. 2014). In this study, IMC was used to evaluate the MBEC of some of the antibiotics previously tested against *S. aureus* #5 and *P. mirabilis* #5 by AB assay, (Fig. 5 and Table 3).

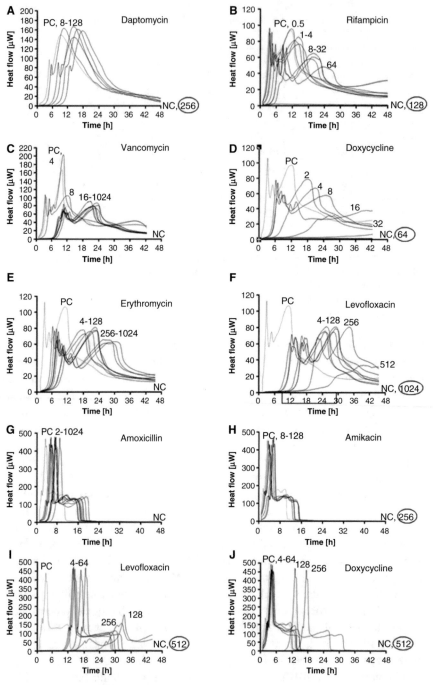

Fig. 5 Determination of the minimal biofilm eradicating concentration (MBEC) by isothermal microcalorimetry. 24 h-old biofilms, grown on porous glass beads, were exposed to different concentrations of various antibiotics. Beads were then washed to remove non-attached bacteria as well as antibiotics and incubated in the microcalorimeter in fresh medium (without antibiotics). The curves represent the heat flow produced over time by *S. aureus* #5 (**A–F**) and *P. mirabilis* #5 (**G–J**). The numbers above each curve indicate the antibiotic concentrations tested expressed in μg/ml. The circled value is the minimum biofilm eradication concentration (MBEC), and corresponds to a flat line indicative of a lack of regrowth. *PC*: positive control, bacterial biofilms without any pre-treatment with antibiotics. *NC*: negative control, medium containing sterile beads

Twenty-four hour-old biofilms grown on porous glass beads were incubated in the presence of two-fold serial dilutions of each antibiotic. Ability of biofilm cells to re-grow after antibiotic treatment was evaluated by IMC incubating beads in antibiotic-free medium. Figure 5 depicts the curves obtained at each antibiotic concentration tested, plotted as heat flow (in microwatt) *versus* time (h). With the exception of vancomycin and erythromycin, which did not show complete inhibition of heat production even at concentrations up to 1024 µg/ml, a complete suppression of heat production was observed in presence of 128 µg/ml rifampicin, 256 µg/ml daptomycin, 64 µg/ml doxycycline, and 1024 µg/ml levofloxacin against *S. aureus* #5 (Fig. 5, Table 3). Amikacin inhibited the heat production by *P. mirabilis* #5 at 256 µg/ml, and levofloxacin and doxycycline at 512 µg/ml. On the contrary, amoxicillin up to 1024 µg/ml, did not abolish heat production by the sessile *P. mirabilis* (Fig. 5, Table 3).

4   Conclusions

Aggregates of bacteria resembling biofilms were detected in the majority of the biopsies of CRS patients. Syto9 staining of the tissue specimens followed by CLSM analysis revealed to be an easy to perform and rapid technique with the potential to be employed in adequately equipped routine microbiology laboratory. The use of more detailed analyses, such as the Z-stack technique, helped to visualise biofilms not only on the tissue surface, but also in the deeper layers of the biopsy, increasing the detection sensitivity. The sampling of multiple biopsies from various locations of the inflamed mucosa could further increase the probability to detect the presence of biofilms.

Interestingly, biofilm-like structures were not detected in any of the control patients, although bacterial strains potentially part of the normal flora were also isolated from biopsies of such subjects. This may suggest that the mere presence of bacteria in the sinus mucosa does not necessarily leads to pathology. It might be possible that in individuals with specific anatomical features or other predisposing factors bacteria originating from the upper respiratory tract or oral cavity may stably colonize the sinus mucosa and form biofilms, thus contributing to the chronicisation of the infection.

The consistent demonstration of biofilms on mucosa biopsies in CRS patients suggests that they may play a role in the pathogenesis/persistence of the disease.

Isolation of biofilm-forming bacteria from biopsies may be troublesome and require appropriate and standardized procedures such as inoculation in an enrichment broth or sonication to allow bacteria to detach from the tissue. Most of the bacterial species isolated from biopsies of CRS patients, belonged to the normal flora of the upper-respiratory tract or oral cavity, while in few cases the isolated bacteria belonged to the *Enterobacteriaceae*.

The type of medium used in the biofilm-forming assay *in vitro* had an impact on the ability to form biofilm of bacterial strains isolated from biopsies of CRS patients and controls. Overall, rich media, resembling the *in vivo* milieu, and containing high glucose concentrations, human plasma, and/or salts, promoted biofilm formation, but not for all the strains. When tested in their optimized growth conditions the large majority of the strains isolated from biopsy samples were moderate or strong biofilm producers *in vitro*.

When a biofilm-associated infection is suspected, antibiotic susceptibility testing against biofilms of the clinical strains isolated from the infection site could be more appropriate than the standard assays performed on planktonic bacteria, to guide the therapy. Nevertheless, standardization of the procedures, parameters and breakpoints, by official agencies, is needed before they are implemented in the clinical microbiology laboratories for routine susceptibility testing.

The introduction of innovative techniques such as isothermal microcalorimetry in the microbiology laboratory may help to perform large scale and accurate antibiotic susceptibility testing against bacteria in biofilms.

Acknowledgments Authors thank Dr. Ranieri Bizzarri for the helpful discussion related to the microscopy analysis. This work was supported by funds from University of Pisa and Consiglio Nazionale delle Ricerche.

References

Abou-Hamad W, Matar N, Elias M, Nasr M, Sarkis-Karam D, Hokayem N, Haddad A (2009) Bacterial flora in normal adult maxillary sinuses. Am J Rhinol Allergy 23(3):261–263. doi:10.2500/ajra.2009.23. 3317

Aly R, Levit S (1987) Adherence of Staphylococcus aureus to squamous epithelium: role of fibronectin and teichoic acid. Rev Infect Dis 9(Suppl 4):S341–S350

Arciola CR, Campoccia D, Gamberini S, Cervellati M, Donati E, Montanaro L (2002) Detection of slime production by means of an optimised Congo red agar plate test based on a colourimetric scale in Staphylococcus epidermidis clinical isolates genotyped for ica locus. Biomaterials 23 (21):4233–4239. doi:Pii S0142–9612(02)00171–0. doi:10.1016/S0142-9612(02)00171-0

Arciola CR, Campoccia D, Gamberini S, Rizzi S, Donati ME, Baldassarri L, Montanaro L (2004) Search for the insertion element IS256 within the ica locus of Staphylococcus epidermidis clinical isolates collected from biomaterial-associated infections. Biomaterials 25 (18):4117–4125. doi:10.1016/j.biomaterials.2003.11. 027

Arciola CR, Campoccia D, Ravaioli S, Montanaro L (2015) Polysaccharide intercellular adhesin in biofilm: structural and regulatory aspects. Front Cell Infect Mi 5. doi:ARTN 7. 10.3389/fcimb.2015.00007

Aslam S, Darouiche RO (2007) Prolonged bacterial exposure to minocycline/rifampicin-impregnated vascular catheters does not affect antimicrobial activity of catheters. J Antimicrob Chemother 60(1):148–151. doi:10.1093/jac/dkm173

Batoni G, Maisetta G, Esin S (2016) Antimicrobial peptides and their interaction with biofilms of medically relevant bacteria. Biochim Biophys Acta 1858 (5):1044–1060. doi:10.1016/j.bbamem.2015.10.013

Benninger MS, Ferguson BJ, Hadley JA, Hamilos DL, Jacobs M, Kennedy DW, Lanza DC, Marple BF, Osguthorpe JD, Stankiewicz JA, Anon J, Denneny J, Emanuel I, Levine H (2003) Adult chronic rhinosinusitis: definitions, diagnosis, epidemiology, and pathophysiology. Otolaryngol Head Neck Surg 129(3 Suppl):S1–32

Bjarnsholt T, Alhede M, Alhede M, Eickhardt-Sorensen SR, Moser C, Kuhl M, Jensen PO, Hoiby N (2013) The in vivo biofilm. Trends Microbiol 21(9):466–474. doi:10.1016/j.tim.2013.06.002

Braissant O, Wirz D, Gopfert B, Daniels AU (2010) Use of isothermal microcalorimetry to monitor microbial activities. FEMS Microbiol Lett 303(1):1–8. doi:10. 1111/j.1574-6968.2009.01819.x

Brancatisano FL, Maisetta G, Di Luca M, Esin S, Bottai D, Bizzarri R, Campa M, Batoni G (2014) Inhibitory effect of the human liver-derived antimicrobial peptide hepcidin 20 on biofilms of polysaccharide intercellular adhesin (PIA)-positive and PIA-negative strains of Staphylococcus epidermidis. Biofouling 30 (4):435–446. doi:10.1080/08927014.2014.888062

Brook I (1981) Bacteriologic features of chronic sinusitis in children. JAMA 246(9):967–969

Brook I (2006) Bacteriology of chronic sinusitis and acute exacerbation of chronic sinusitis. Arch Otolaryngol Head Neck Surg 132(10):1099–1101. doi:10.1001/archotol.132.10.1099

Brook I (2011) Microbiology of sinusitis. Proc Am Thorac Soc 8(1):90–100. doi:10.1513/pats.201006-038RN

Brook I, Frazier EH (2001) Correlation between microbiology and previous sinus surgery in patients with chronic maxillary sinusitis. Ann Otol Rhinol Laryngol 110(2):148–151

Brooks JL, Jefferson KK (2014) Phase variation of poly-N-acetylglucosamine expression in Staphylococcus aureus. PLoS Pathog 10(7):e1004292. doi:10.1371/journal.ppat.1004292

Cardile AP, Sanchez CJ Jr, Samberg ME, Romano DR, Hardy SK, Wenke JC, Murray CK, Akers KS (2014) Human plasma enhances the expression of Staphylococcal microbial surface components recognizing adhesive matrix molecules promoting biofilm formation and increases antimicrobial tolerance in vitro. BMC Res Notes 7:457. doi:10.1186/1756-0500-7-457

Chen P, Abercrombie JJ, Jeffrey NR, Leung KP (2012) An improved medium for growing Staphylococcus aureus biofilm. J Microbiol Methods 90(2):115–118. doi:10.1016/j.mimet.2012.04.009

Coffey CS, Sonnenburg RE, Melroy CT, Dubin MG, Senior BA (2006) Endoscopically guided aerobic cultures in postsurgical patients with chronic rhinosinusitis. Am J Rhinol 20(1):72–76

Cook HE, Haber J (1987) Bacteriology of the maxillary sinus. J Oral Maxillofac Surg 45(12):1011–1014

Costerton JW, Post JC, Ehrlich GD, Hu FZ, Kreft R, Nistico L, Kathju S, Stoodley P, Hall-Stoodley L, Maale G, James G, Sotereanos N, DeMeo P (2011) New methods for the detection of orthopedic and other biofilm infections. FEMS Immunol Med Microbiol 61(2):133–140. doi:10.1111/j.1574-695X. 2010.00766.x

Cue D, Lei MG, Luong TT, Kuechenmeister L, Dunman PM, O'Donnell S, Rowe S, O'Gara JP, Lee CY (2009) Rbf promotes biofilm formation by Staphylococcus Aureus via repression of icaR, a negative regulator of icaADBC. J Bacteriol 191(20):6363–6373. doi:10. 1128/jb.00913-09

Danielsen KA, Eskeland O, Fridrich-Aas K, Orszagh VC, Bachmann-Harildstad G, Burum-Auensen E (2014) Bacterial biofilms in patients with chronic rhinosinusitis: a confocal scanning laser microscopy study. Rhinology 52(2):150–155. doi:10.4193/Rhin

Desrosiers M, Evans GA, Keith PK, Wright ED, Kaplan A, Bouchard J, Ciavarella A, Doyle PW, Javer AR, Leith ES, Mukherji A, Robert Schellenberg R, Small P, Witterick IJ (2011) Canadian clinical practice guidelines for acute and chronic rhinosinusitis. J Otolaryngol Head Neck Surg 40 (Suppl 2):S99–193

Di Domenico EG, Toma L, Provot C, Ascenzioni F, Sperduti I, Prignano G, Gallo MT, Pimpinelli F, Bordignon V, Bernardi T, Ensoli F (2016) Development of an in vitro assay, based on the BioFilm ring test(R), for rapid profiling of biofilm-growing bacteria. Front Microbiol 7:1429. doi:10.3389/fmicb.2016. 01429

Donlan RM, Costerton JW (2002) Biofilms: survival mechanisms of clinically relevant microorganisms. Clin Microbiol Rev 15(2):167–193

Dunne WM Jr, Mason EO Jr, Kaplan SL (1993) Diffusion of rifampin and vancomycin through a Staphylococcus epidermidis biofilm. Antimicrob Agents Chemother 37(12):2522–2526

Fey PD (2010) Modality of bacterial growth presents unique targets: how do we treat biofilm-mediated infections? Curr Opin Microbiol 13(5):610–615. doi:10.1016/j.mib.2010.09.007

Fokkens WJ, Lund VJ, Mullol J, Bachert C, Alobid I, Baroody F, Cohen N, Cervin A, Douglas R, Gevaert P, Georgalas C, Goossens H, Harvey R, Hellings P, Hopkins C, Jones N, Joos G, Kalogjera L, Kern B, Kowalski M, Price D, Riechelmann H, Schlosser R, Senior B, Thomas M, Toskala E, Voegels R, de Wang Y, Wormald PJ (2012) European Position Paper on Rhinosinusitis and Nasal Polyps 2012. Rhinol Suppl (23):3 p preceding table of contents, 1–298

Foreman A, Psaltis AJ, Tan LW, Wormald PJ (2009) Characterization of bacterial and fungal biofilms in chronic rhinosinusitis. Am J Rhinol Allergy 23 (6):556–561. doi:10.2500/ajra.2009.23.3413

Foreman A, Singhal D, Psaltis AJ, Wormald PJ (2010) Targeted imaging modality selection for bacterial biofilms in chronic rhinosinusitis. Laryngoscope 120 (2):427–431. doi:10.1002/lary.20705

Freeman DJ, Falkiner FR, Keane CT (1989) New method for detecting slime production by coagulase negative staphylococci. J Clin Pathol 42(8):872–874

Furustrand Tafin U, Meis JF, Trampuz A (2012) Isothermal microcalorimetry for antifungal susceptibility testing of Mucorales, Fusarium spp., and Scedosporium spp. Diagn Microbiol Infect Dis 73 (4):330–337. doi:10.1016/j.diagmicrobio.2012.05.009

Furustrand Tafin U, Orasch C, Trampuz A (2013) Activity of antifungal combinations against Aspergillus species evaluated by isothermal microcalorimetry. Diagn Microbiol Infect Dis 77(1):31–36. doi:10.1016/j. diagmicrobio.2013.06.004

Gil C, Solano C, Burgui S, Latasa C, Garcia B, Toledo-Arana A, Lasa I, Valle J (2014) Biofilm matrix exoproteins induce a protective immune response against Staphylococcus aureus biofilm infection. Infect Immun 82(3):1017–1029. doi:10.1128/iai. 01419-13

Givskov M, Ostling J, Eberl L, Lindum PW, Christensen AB, Christiansen G, Molin S, Kjelleberg S (1998) Two separate regulatory systems participate in control of swarming motility of Serratia liquefaciens MG1. J Bacteriol 180(3):742–745

Hall-Stoodley L, Stoodley P, Kathju S, Hoiby N, Moser C, Costerton JW, Moter A, Bjarnsholt T (2012) Towards diagnostic guidelines for biofilm-associated infections. FEMS Immunol Med Microbiol 65(2):127–145. doi:10.1111/j.1574-695X.2012. 00968.x

Hoiby N, Bjarnsholt T, Moser C, Bassi GL, Coenye T, Donelli G, Hall-Stoodley L, Hola V, Imbert C, Kirketerp-Moller K, Lebeaux D, Oliver A, Ullmann AJ, Williams C, Biofilms ESGf, Consulting External Expert, Werner Z (2015) ESCMID guideline for the diagnosis and treatment of biofilm infections 2014. Clin Microbiol Infect 21(Suppl 1):S1–25. doi:10. 1016/j.cmi.2014.10.024

Jefferson KK, Goldmann DA, Pier GB (2005) Use of confocal microscopy to analyze the rate of vancomycin penetration through Staphylococcus aureus biofilms. Antimicrob Agents Chemother 49 (6):2467–2473. doi:10.1128/aac.49.6.2467-2473.2005

Jiang RS, Liang KL, Jang JW, Hsu CY (1999) Bacteriology of endoscopically normal maxillary sinuses. J Laryngol Otol 113(9):825–828

Jost GF, Wasner M, Taub E, Walti L, Mariani L, Trampuz A (2014) Sonication of catheter tips for improved detection of microorganisms on external ventricular drains and ventriculo-peritoneal shunts. J Clin Neurosci 21(4):578–582. doi:10.1016/j.jocn.2013.05. 025

Kim R, Freeman J, Waldvogel-Thurlow S, Roberts S, Douglas R (2013a) The characteristics of intramucosal bacteria in chronic rhinosinusitis: a prospective cross-sectional analysis. Int Forum Allergy Rhinol 3 (5):349–354. doi:10.1002/alr.21117

Kim RJ, Yin T, Chen CJ, Mansell CJ, Wood A, Dunbar PR, Douglas RG (2013b) The interaction between bacteria and mucosal immunity in chronic rhinosinusitis: a prospective cross-sectional analysis. Am J Rhinol Allergy 27(6):e183–e189. doi:10.2500/ ajra.2013.27.3974

Kim YM, Jin J, Choi JA, Cho SN, Lim YJ, Lee JH, Seo JY, Chen HY, Rha KS, Song CH (2014) Staphylococcus aureus enterotoxin B-induced endoplasmic reticulum stress response is associated with chronic rhinosinusitis with nasal polyposis. Clin Biochem 47(1–2):96–103. doi:10.1016/j.clinbiochem.2013.10.030

Knobloch JK, Bartscht K, Sabottke A, Rohde H, Feucht HH, Mack D (2001) Biofilm formation by

Staphylococcus epidermidis depends on functional RsbU, an activator of the sigB operon: differential activation mechanisms due to ethanol and salt stress. J Bacteriol 183(8):2624–2633. doi:10.1128/JB.183.8.2624-2633.2001

Lam K, Schleimer R, Kern RC (2015) The etiology and pathogenesis of chronic Rhinosinusitis: a review of current hypotheses. Curr Allergy Asthma Rep 15 (7):41. doi:10.1007/s11882-015-0540-2

Lanza DC, Kennedy DW (1997) Adult rhinosinusitis defined. Otolaryngol Head Neck Surg 117(3 Pt 2): S1–S7

Lebeaux D, Ghigo JM, Beloin C (2014) Biofilm-related infections: bridging the gap between clinical management and fundamental aspects of recalcitrance toward antibiotics. Microbiol Mol Biol Rev 78(3):510–543. doi:10.1128/MMBR.00013-14

Lewis K (2008) Multidrug tolerance of biofilms and persister cells. Curr Top Microbiol Immunol 322:107–131

Lim Y, Jana M, Luong TT, Lee CY (2004) Control of glucose- and NaCl-induced biofilm formation by rbf in Staphylococcus aureus. J Bacteriol 186(3):722–729

Lubbert C, Wendt K, Feisthammel J, Moter A, Lippmann N, Busch T, Mossner J, Hoffmeister A, Rodloff AC (2016) Epidemiology and resistance patterns of bacterial and fungal colonization of biliary plastic stents: a prospective cohort study. PLoS One 11(5):e0155479. doi:10.1371/journal.pone.0155479

Macia MD, Rojo-Molinero E, Oliver A (2014) Antimicrobial susceptibility testing in biofilm-growing bacteria. Clin Microbiol Infect 20(10):981–990. doi:10.1111/1469-0691.12651

Maisetta G, Grassi L, Di Luca M, Bombardelli S, Medici C, Brancatisano FL, Esin S, Batoni G (2016) Anti-biofilm properties of the antimicrobial peptide temporin 1Tb and its ability, in combination with EDTA, to eradicate Staphylococcus epidermidis biofilms on silicone catheters. Biofouling 32 (7):787–800. doi:10.1080/08927014.2016.1194401

Mallmann C, Siemoneit S, Schmiedel D, Petrich A, Gescher DM, Halle E, Musci M, Hetzer R, Gobel UB, Moter A (2010) Fluorescence in situ hybridization to improve the diagnosis of endocarditis: a pilot study. Clin Microbiol Infect 16(6):767–773. doi:10.1111/j.1469-0691.2009.02936.x

Martin-Rodriguez AJ, Gonzalez-Orive A, Hernandez-Creus A, Morales A, Dorta-Guerra R, Norte M, Martin VS, Fernandez JJ (2014) On the influence of the culture conditions in bacterial antifouling bioassays and biofilm properties: Shewanella algae, a case study. BMC Microbiol 14:102. doi:10.1186/1471-2180-14-102

Merino N, Toledo-Arana A, Vergara-Irigaray M, Valle J, Solano C, Calvo E, Lopez JA, Foster TJ, Penades JR, Lasa I (2009) Protein A-mediated multicellular behavior in Staphylococcus Aureus. J Bacteriol 191 (3):832–843. doi:10.1128/jb.01222-08

Mladina R, Skitarelic N, Music S, Ristic M (2010) A biofilm exists on healthy mucosa of the paranasal sinuses: a prospectively performed, blinded, scanning electron microscope study. Clin Otolaryngol 35 (2):104–110. doi:10.1111/j.1749-4486.2010.02097.x

Oliva A, Furustrand Tafin U, Maiolo EM, Jeddari S, Betrisey B, Trampuz A (2014) Activities of fosfomycin and rifampin on planktonic and adherent Enterococcus faecalis strains in an experimental foreign-body infection model. Antimicrob Agents Chemother 58(3):1284–1293. doi:10.1128/AAC.02583-12

Paes Leme AF, Koo H, Bellato CM, Bedi G, Cury JA (2006) The role of sucrose in cariogenic dental biofilm formation--new insight. J Dent Res 85(10):878–887

Penesyan A, Gillings M, Paulsen IT (2015) Antibiotic discovery: combatting bacterial resistance in cells and in biofilm communities. Molecules 20 (4):5286–5298. doi:10.3390/molecules20045286

Peters AT, Spector S, Hsu J, Hamilos DL, Baroody FM, Chandra RK, Grammer LC, Kennedy DW, Cohen NA, Kaliner MA, Wald ER, Karagianis A, Slavin RG, Joint Task Force on Practice Parameters rtAAoAA, Immunology tACoAA, Immunology, the Joint Council of Allergy A, Immunology (2014) Diagnosis and management of rhinosinusitis: a practice parameter update. Ann Allergy Asthma Immunol 113 (4):347–385. doi:10.1016/j.anai.2014.07.025

Pettit RK, Weber CA, Kean MJ, Hoffmann H, Pettit GR, Tan R, Franks KS, Horton ML (2005) Microplate Alamar blue assay for Staphylococcus epidermidis biofilm susceptibility testing. Antimicrob Agents Chemother 49(7):2612–2617. doi:10.1128/AAC.49.7.2612-2617.2005

Portillo ME, Salvado M, Trampuz A, Siverio A, Alier A, Sorli L, Martinez S, Perez-Prieto D, Horcajada JP, Puig-Verdie L (2015) Improved diagnosis of orthopedic implant-associated infection by inoculation of sonication fluid into blood culture bottles. J Clin Microbiol 53(5):1622–1627. doi:10.1128/JCM.03683-14

Psaltis AJ, Ha KR, Beule AG, Tan LW, Wormald PJ (2007) Confocal scanning laser microscopy evidence of biofilms in patients with chronic rhinosinusitis. Laryngoscope 117(7):1302–1306. doi:10.1097/MLG.0b013e31806009b0

Raad II, Darouiche RO, Hachem R, Abi-Said D, Safar H, Darnule T, Mansouri M, Morck D (1998) Antimicrobial durability and rare ultrastructural colonization of indwelling central catheters coated with minocycline and rifampin. Crit Care Med 26(2):219–224

Ravaioli S, Selan L, Visai L, Pirini V, Campoccia D, Maso A, Speziale P, Montanaro L, Arciola CR (2012) Staphylococcus lugdunensis, an aggressive coagulase-negative pathogen not to be underestimated. Int J Artif Organs 35(10):742–753. doi:10.5301/ijao.5000142

Richter AL, Gallagher KK (2016) Chronic invasive fungal sinusitis causing a pathologic Le Fort I fracture in an immunocompetent patient. Ear Nose Throat J 95 (9):E1–E3

Rosenfeld RM, Piccirillo JF, Chandrasekhar SS, Brook I, Ashok Kumar K, Kramper M, Orlandi RR, Palmer JN, Patel ZM, Peters A, Walsh SA, Corrigan MD (2015)

Clinical practice guideline (update): adult sinusitis. Otolaryngol Head Neck Surg 152(2 Suppl):S1–S39. doi:10.1177/0194599815572097

Saginur R, Stdenis M, Ferris W, Aaron SD, Chan F, Lee C, Ramotar K (2006) Multiple combination bactericidal testing of staphylococcal biofilms from implant-associated infections. Antimicrob Agents Chemother 50(1):55–61. doi:10.1128/aac.50.1.55-61. 2006

Sanchez CJ Jr, Mende K, Beckius ML, Akers KS, Romano DR, Wenke JC, Murray CK (2013) Biofilm formation by clinical isolates and the implications in chronic infections. BMC Infect Dis 13:47. doi:10. 1186/1471-2334-13-47

Shields RC, Mokhtar N, Ford M, Hall MJ, Burgess JG, ElBadawey MR, Jakubovics NS (2013) Efficacy of a marine bacterial nuclease against biofilm forming microorganisms isolated from chronic rhinosinusitis. PLoS One 8(2):e55339. doi:10.1371/journal.pone. 0055339

Singhal D, Psaltis AJ, Foreman A, Wormald PJ (2010) The impact of biofilms on outcomes after endoscopic sinus surgery. Am J Rhinol Allergy 24(3):169–174. doi:10.2500/ajra.2010.24.3462

Snyman J, Claassen AJ, Botha PL (1988) A microbiological study of acute maxillary sinusitis in Bloemfontein. S Afr Med J 74(9):444–445

Sobin J, Engquist S, Nord CE (1992) Bacteriology of the maxillary sinus in healthy volunteers. Scand J Infect Dis 24(5):633–635

Speziale P, Pietrocola G, Foster TJ, Geoghegan JA (2014) Protein-based biofilm matrices in staphylococci. Front Cell Infect Microbiol 4:171. doi:10.3389/fcimb.2014. 00171

Stepanovic S, Vukovic D, Hola V, Di Bonaventura G, Djukic S, Cirkovic I, Ruzicka F (2007) Quantification of biofilm in microtiter plates: overview of testing conditions and practical recommendations for assessment of biofilm production by staphylococci. APMIS 115(8):891–899. doi:10.1111/j.1600-0463.2007.apm_ 630.x

Sun F, Qu F, Ling Y, Mao P, Xia P, Chen H, Zhou D (2013) Biofilm-associated infections: antibiotic resistance and novel therapeutic strategies. Future Microbiol 8(7):877–886. doi:10.2217/fmb.13.58

Tote K, Vanden Berghe D, Maes L, Cos P (2008) A new colorimetric microtitre model for the detection of Staphylococcus aureus biofilms. Lett Appl Microbiol 46(2):249–254. doi:10.1111/j.1472-765X.2007. 02298.x

Trampuz A, Zimmerli W (2006) Antimicrobial agents in orthopaedic surgery: prophylaxis and treatment. Drugs 66(8):1089–1105

Trampuz A, Piper KE, Jacobson MJ, Hanssen AD, Unni KK, Osmon DR, Mandrekar JN, Cockerill FR, Steckelberg JM, Greenleaf JF, Patel R (2007) Sonication of removed hip and knee prostheses for diagnosis of infection. N Engl J Med 357(7):654–663. doi:10. 1056/NEJMoa061588

Vaudaux PE, Huggler E, Lerch PG, Morgenthaler JJ, Nydegger UE, Schumacher-Perdreau F, Lew PD, Waldvogel FA (1989) Inhibition by immunoglobulins of Staphylococcus aureus adherence to fibronectin-coated foreign surfaces. J Invest Surg 2(4):397–408

Vergara-Irigaray M, Valle J, Merino N, Latasa C, Garcia B, Ruiz de Los Mozos I, Solano C, Toledo-Arana A, Penades JR, Lasa I (2009) Relevant role of fibronectin-binding proteins in Staphylococcus aureus biofilm-associated foreign-body infections. Infect Immun 77(9):3978–3991. doi:10.1128/iai.00616-09

Wagner C, Aytac S, Hansch GM (2011) Biofilm growth on implants: bacteria prefer plasma coats. Int J Artif Organs 34(9):811–817. doi:10.5301/ijao.5000061

Walker JN, Horswill AR (2012) A coverslip-based technique for evaluating Staphylococcus aureus biofilm formation on human plasma. Front Cell Infect Microbiol 2:39. doi:10.3389/fcimb.2012.00039

Wecke J, Kersten T, Madela K, Moter A, Gobel UB, Friedmann A, Bernimoulin J (2000) A novel technique for monitoring the development of bacterial biofilms in human periodontal pockets. FEMS Microbiol Lett 191(1):95–101

Widmer AF, Gaechter A, Ochsner PE, Zimmerli W (1992) Antimicrobial treatment of orthopedic implant-related infections with rifampin combinations. Clin Infect Dis 14(6):1251–1253

Wood AJ, Fraser JD, Swift S, Patterson-Emanuelson EA, Amirapu S, Douglas RG (2012) Intramucosal bacterial microcolonies exist in chronic rhinosinusitis without inducing a local immune response. Am J Rhinol Allergy 26(4):265–270. doi:10.2500/ajra.2012.26.3779

Zheng Z, Stewart PS (2002) Penetration of rifampin through Staphylococcus epidermidis biofilms. Antimicrob Agents Chemother 46(3):900–903

Ziebuhr W, Krimmer V, Rachid S, Lossner I, Gotz F, Hacker J (1999) A novel mechanism of phase variation of virulence in Staphylococcus epidermidis: evidence for control of the polysaccharide intercellular adhesin synthesis by alternating insertion and excision of the insertion sequence element IS256. Mol Microbiol 32(2):345–356

Zimmerli W, Widmer AF, Blatter M, Frei R, Ochsner PE (1998) Role of rifampin for treatment of orthopedic implant-related staphylococcal infections: a randomized controlled trial. Foreign-body infection (FBI) study group. JAMA 279(19):1537–1541

Adv Exp Med Biol - Advances in Microbiology, Infectious Diseases and Public Health (2018) 9: 29–39
DOI 10.1007/5584_2017_40
© Springer International Publishing AG 2017
Published online: 27 April 2017

Immune Response Against *Salmonella* Enteritidis Is Unsettled by HIV Infection

Maira Costa Cacemiro, Milena Sobral Espíndola,
Leonardo Judson Galvão-Lima, Luana Silva Soares,
Caroline Fontanari, Marco Aurélio Prata, Fábio Campioni,
Juliana Pfrimer Falcão, Valdes Roberto Bollela,
and Fabiani Gai Frantz

Abstract

The human immunodeficiency virus (HIV) is responsible for more than 2 million new infections per year and opportunistic infections such as *Salmonella spp.* Gastroenteritis is an important cause of mortality and morbidity in developing countries. Monocytes and macrophages play a critical role in the innate immune response against bacterial infections. However during HIV infection the virus can infect these cells and although they are more resistant to the cytopathic effects, they represent an important viral reservoir in these patients. Our aim was to evaluate the monocyte functions from HIV-1 infected patients after *in vitro* exposition to *Salmonella* Enteritidis. Our results suggest impairment of monocytes phagocytic and microbicidal activity in HIV-1 non-treated patients, which was more evident in women, if compared with men. Moreover, monocytes from HIV-1 infected and non-treated patients after stimulation with the bacteria, produced more pro-inflammatory cytokines than monocytes from HIV-treated patients, suggesting that HIV-1 infected patients have their functions unbalanced, once in the presence of an opportunistic infection *in vitro*.

Keywords

HIV infection · Monocytes · Antiretroviral therapy · *Salmonella enterica* serotype Enteritidis

M.C. Cacemiro, M.S. Espíndola, L.J. Galvão-Lima,
L.S. Soares, C. Fontanari, F. Campioni, J.P. Falcão,
and F.G. Frantz (✉)
Department of Clinical Analysis, Toxicology, and Food Sciences, School of Pharmaceutical Sciences of Ribeirão Preto, University of São Paulo – USP, Ribeirão Preto, SP, Brazil
e-mail: mairacacemiro@usp.br;
milenasespindola@gmail.com; galvaolima@usp.br;
luanasisoares@gmail.com; fontanari@usp.br;
fabiocampioni@yahoo.com; jufalcao@fcfrp.usp.br;
frantz@usp.br

M.A. Prata
Genetic Department – Ribeirão Preto Medical School, University of São Paulo – USP, Ribeirão Preto, SP, Brazil
e-mail: marcoaurelioprata@yahoo.com.br

V.R. Bollela
Department of Internal Medicine, Ribeirão Preto Medical School, University of São Paulo – USP, Ribeirão Preto, SP, Brazil
e-mail: bollela@gmail.com

1   Introduction

In 1983 the human immunodeficiency virus (HIV) was identified by two independent groups headed by Robert Gallo and Luc Montagnier (Barré-Sinoussi et al. 1983; Gorry et al. 2005). Since then, millions of people are still dying of AIDS. According to UNAIDS, 2.1 million of new infections was reported in 2015, and it was estimated that 36.7 million of people were infected by HIV (UNAIDS 2016). The Highly Active Antiretroviral Treatment (HAART) adhesion is directly related to the increase in the life expectancy attributed to the reduction in deaths due to the progression to AIDS and consequently to a drastic reduction in the number of co-opportunistic infections (Thompson et al. 2012). However, in developing countries where therapy is not fully accessible, opportunistic infections are still a significant cause of morbidity and mortality (Dhanoa and Fatt 2009), mainly in people who are not under HAART. Gastroenteritis caused by *Salmonella* spp. is among the common opportunistic diseases correlated with HIV infection and may cause a recurrent septicemia in these patients (Bhagani 2011; Clumeck et al. 1983; MacLennan et al. 2010; Preziosi et al. 2012).

During bacterial infections in general, monocytes and macrophages play a critical role in the innate immune response, so monocytes and macrophages should be able to prevent bacteria from causing opportunistic infections (Kuroda 2010). However, during HIV infection, macrophages are the first target of the virus, which are more resistant to the cytopathic effects, becoming an important viral reservoir (Kuroda 2010; Meléndez et al. 2011). In this way, it was postulated that HIV infection could interfere in cellular functions, being the reason of opportunistic pathogens escaping from the host defense (Kuroda 2010; Zybarth et al. 1999). Our group and others has shown that HIV-1 infected patients have a severe inflammatory state and that even after HAART and reduction in viremia,

these individuals continue to have an important immune dysregulation (Burdo et al. 2011; Espíndola et al. 2015).

Moreover, differential susceptibility to infections between women and men have been documented in the literature for a variety of pathogens, including HIV infection (Klein 2012). Gender-based differences in disease comprise genetic differences, as well as expression of steroid hormones, differences in anatomy, and X and Y chromosome–linked factors. It has been well established through cross-sectional and longitudinal studies in HIV-1 infection that in acute and chronic HIV-1 infection, women have lower plasma viral loads and higher $CD4^+$ T-cell counts than men and despite harboring lower viral loads for a given level of viremia, women had a 1.6-fold higher risk of progressing to AIDS (Farzadegan et al. 1998; Meditz et al. 2011; Sterling et al. 1999; Touloumi et al. 2004).

Therefore, our aim was to identify possible functional changes in monocytes from HIV-1 infected patients, when these cells are infected *in vitro* with the opportunistic *Salmonella* Enteritidis. Here we showed that monocytes from HIV-1 patients suffer the effects of the inflammatory dysregulation and when cultured in the presence of *S.* Enteritidis, were not able to properly phagocytize and kill the bacteria, and consequently present dysregulated cytokines production, beyond that, we also showed lower phagocytic and microbicidal activity and an augment in cytokines/chemokines production by women compared to men.

2   Ethical Approval

This work was approved by the Ethics Committee from Faculdade de Ciências Farmacêuticas de Ribeirão Preto - Universidade de São Paulo - USP (Protocol n° 216), as well as by Fundação Hemocentro de Ribeirão Preto and by Hospital das Clínicas de Ribeirão Preto. All patients and healthy donors signed an informed written consent form in accordance with the guidelines established by the Brazilian National Health Council.

3 Casuistry and Methods

The non-infected subjects were selected at Fundação Hemocentro de Ribeirão Preto aged between 18 and 65 years, from both genders, were HIV-1 negative and had no history of chronic illness or drug use. All HIV-1 infected patients were selected from the special unit for HIV patient care at Hospital das Clínicas, School of Medicine of Ribeirão Preto, University of São Paulo – Brazil. Twenty subjects without HIV infection composed the control group (CTRL). The group of non-treated HIV patients (HIV-NT) was composed by 13 HIV-1-infected subjects who were not receiving antiretroviral treatment. The HIV-treated (HIV-T) group was composed by 8 subjects under regular antiretroviral treatment, over 6 months, and with undetectable (<40 copies/mL) HIV viral load. HIV+ patients from both genders were recruited and aged between 18 and 65 years. The characteristics of the study group such as gender, age, $CD4^+$ and $CD8^+$ T cell number and the viral load are shown in Table 1.

3.1 Blood Samples

Peripheral blood (40 mL) was collected from each individual in vacuum tubes containing

Table 1 Characteristics of the study population

Groups	Age	$CD4^+$ cell/mm^3	$CD8^+$ cell/mm^3	Viral load copies/mL
CTRL	36	–	–	–
HIV-NT	37	590.46	874.38	32,913
HIV-T	45	497.5	1119.91	<40–

The values represent the median of each group
CTRL control, *HIV-NT* HIV non-treated, *HIV-T* HIV in regular antiretroviral therapy

heparin. The blood was centrifuged at $400 \times g$ (Thermo Scientific – Megafuge 16R Centrifuge) for 10 min in room temperature to remove the plasma.

3.2 Peripheral Blood Mononuclear Cells (PBMC)

The blood without plasma, was mixed with phosphate buffered saline (PBS) and added over Ficoll-Paque®PLUS d = 1078 g/mL (GE Healthcare Bio-Sciences AB, Uppsala, Sweden). Then was centrifuged at $600 \times g$ for 30 min in room temperature, forming a ring of mononuclear cells that were collected and washed twice with PBS at $400 \times g$ for 10 min in room temperature.

3.3 Monocytes Separation by Magnetic Microbeads

The PBMC were resuspended in a buffer (PBS $1\times$ free of Ca^{2+} e Mg^{2+} + 2 mM of EDTA + 0.5% serum bovine fetal) and centrifuged at $300 \times g$ for 10 min at 4 °C. The cell precipitate was resuspended in buffer and anti-CD14 antibody conjugated with magnetic microbeads was added to the cell suspension. After incubation, cells were washed with buffer and centrifuged for 10 min at $300 \times g$ at 4 °C. The pellets were resuspended in buffer and applied in a magnetic column for achievement of CD14+ cells.

3.4 *Salmonella* Enteritidis

The strain of *Salmonella enterica* subspecies *enterica* serovarity Enteritidis used was a courtesy of the Laboratory of Molecular Epidemiology, Virulence and Bacterial Genomics at University of São Paulo – School of Pharmaceutical Sciences of Ribeirão Preto. The strains were previously described by Campioni and colleagues (Campioni et al. 2012).

The bacteria were grown in brain heart infusion broth (BHI) medium (Himedia) at 37 °C for 5 h with 5% CO_2. The multiplicity of infection (MOI) was calculated taking into account the colony forming units (CFU), obtained after bacteria growth in Mueller Hinton agar (Acumedia – Neogen Corporation). For the experiments aiming the *in vitro* monocytes stimulation, the *S.* Enteritidis were heat-killed (121 °C) for 30 min.

3.5 Phagocytosis

Monocytes (CD14+ cells) were adjusted to a suspension of 1×10^5 cells/mL, cultivated in 96 wells plate in supplemented RPMI-1640 (Gibco) (0.01 mg/mL of penicillin and strepto-mycin (Gibco) and 10% of fetal bovine serum) and incubated overnight at 37 °C and 5% CO_2. After incubation, the medium was removed and *S.* Enteritidis were added (MOI 50) in RPMI-1640 medium supplemented with 50 µg/mL of chloramphenicol. Thirty minutes after incubation of the cells with the bacteria, the plate was washed twice with PBS to remove non-phagocytized bacteria. The cells were lysed with Saponin (Quilaja Bark – Sigma) 0.05%, allowing phagocytized bacteria to be found on the supernatant and 50 µg/mL of resazurin (Resazurin Sodium Salt – Sigma Aldrich) was added to the supernatant. The plate was incubated for additional 7 h at 37 °C and 5% CO_2 in the absence of light, allowing the resazurin metabolization by phagocytized bacteria. The fluorescence of resazurin was evaluated on a microplate fluorometer (SpectraMax Paradigm Multi-mode detection platform, Molecular Devices) with excitation length of 590 nm and emission length of 560 nm.

3.6 Microbicidal Activity

The protocol used was the same as described in the phagocytosis assay, however, after 30 min of incubation of the bacteria with monocytes, the plate was washed to remove non-phagocytized

bacteria and incubated again at the same conditions for additional 90 min to allow the monocytes to perform the microbicidal activity. The cells were lysed with Saponin (Quilaja Bark – Sigma) 0.05%, allowing phagocytized bacteria to be quantified in the supernatant by the addition of 50 µg/mL of resazurin (Resazurin Sodium Salt – Sigma Aldrich). The plate was incubated for additional 7 h at 37 °C and 5% CO_2 in the absence of light for resazurin metabolization by the bacteria. The fluorescence of resazurin was evaluated on a microplate fluorometer (SpectraMax Paradigm Multi-mode detection platform, Molecular Devices) with excitation length of 590 nm and emission length of 560 nm.

3.7 Cytokines and Chemokines Measurement in the Supernatants

The cytokines: IL-1β, IL-5, IL-6, IL-10, IL-12 (p70), IL-4, IFN-α2, IFN-γ, TNF-α, and the chemokines: IP-10 (CXCL10), RANTES (CCL5) and MCP-1 (CCL2) were quantified in the supernatants of monocytes from 17 CTRL subjects of which 11 men and 6 women, 8 HIV-NT patients of which 6 are men and 2 women and 12 HIV-T patients of which 9 are men and 3 are women. Monocytes were stimulated or not with heat-killed *S.* Enteritidis (MOI 50) for 24 h at 37 °C and 5% CO_2, using a multiplex assay (HCYTOMAG-60K, Milliplex® kit, Merck Millipore, Germany) with Luminex® Magpix™ technology (Austin, TX, USA). The assay was performed following the manufacturing instructions. The cytokines/chemokines concentrations were calculated by Milliplex Analyst 5.1® software using a standard curve with cubic spline fitting (log scale).

4 Statistical Analysis

SAS (2008) was the software used for the statistical analysis. For phagocytosis, microbicidal activity, and ROS analyses, the test used was PROC GLM and the media were compared by Tukey test 5%.

Furthermore, the cytokine profile of monocytes from each individual after bacteria stimulation was assessed. All data of cytokine/chemokine production were used to calculate the global median value, which was considered as a cut-off point to categorize each individual as "low" or "high" producers of a given cytokine/chemokine. Data were organized in black and white scale diagrams to determine the frequency of high producers for each group. Relevant data ($\geq$50%) were highlighted in both bold and underline format. Radar charts were performed in Microsoft Excel (Microsoft Office 2010, Las Vegas, USA) to define the overall frequency of individuals with high levels of each cytokine/chemokine for all study groups.

5 Results

5.1 Impaired Phagocytic and Microbicidal Capacity Among HIV-1 Infected Patients Without Antiretroviral Treatment

The phagocytic and microbicidal activity were evaluated after *Salmonella* Enteritidis infection and there was a lower phagocytic ability in monocytes from HIV-NT, when compared with CTRL and HIV-T groups and lower microbicidal activity from HIV-NT patients compared to HIV-T patients (Table 2).

Furthermore, considering the gender as a variable, we observed an increased phagocytic function of monocytes from men compared with women in both HIV-NT and HIV-T groups (Table 3). Regarding the microbicidal activity we observed increased microbicidal activity of monocytes from men compared with women in HIV-NT group (Table 4).

Table 2 Representation of phagocytosis and microbicidal activity

Groups	Phagocytosis	Microbicidal activity
CTRL	1.26 ± 0.21^a	$0.83 \pm 0.13^{a,b}$
HIV-NT	1.01 ± 0.28^b	0.90 ± 0.16^b
HIV-T	1.73 ± 0.32^a	$1.15 + 0.19^a$

Same letters in the same column do not differ by Tukey test (5.0%)
CTRL control, *HIV-NT* HIV non-treated, *HIV-T* HIV in regular antiretroviral therapy

Table 3 Median per group of phagocytic activity compared between genders

Phagocytosis

	Men	Women
Groups	Mean $\pm$ SE	Mean $\pm$ SE
CTRL	1.73 ± 0.27^a	0.82 ± 0.34^a
HIV-NT	1.66 ± 0.35^a	0.09 ± 0.68^b
HIV-T	2.16 ± 0.36^a	1.34 ± 0.43^b

Same letters in the same row do not differ by Tukey test (5.0%)
CTRL control, *HIV-NT* HIV non-treated, *HIV-T* HIV in regular antiretroviral therapy, *SE* standard error

Table 4 Median of microbicidal activity compared between genders

Microbicidal activity

	Men	Women
Groups	Mean $\pm$ SE	Mean $\pm$ SE
CTRL	0.99 ± 0.13^a	0.71 ± 0.16^a
HIV-NT	1.32 ± 0.25^a	0.07 ± 0.48^b
HIV-T	1.25 ± 0.22^a	1.10 ± 0.26^a

Same letters in the same row do not differ by Tukey test (5.0%)
CTRL control, *HIV-NT* HIV non-treated, *HIV-T* HIV in regular antiretroviral therapy, *SE* standard error

5.2 Monocytes from HIV-1-Infected Treated Patients Showed Lower Production of Cytokines/Chemokines After Stimulation with *S*. Enteritidis

To verify the modulation of HIV-1 infection on cytokines and chemokines production by monocytes after exposition to an opportunistic pathogen, we evaluated the cytokines and chemokines released when monocytes were

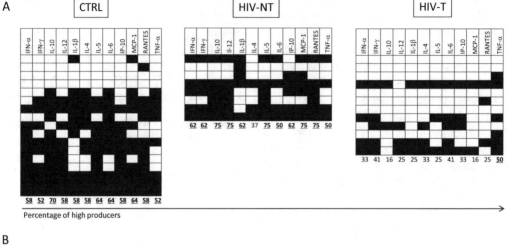

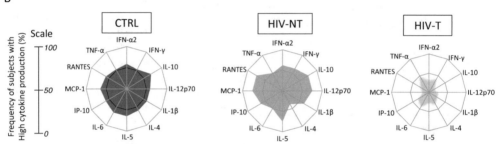

Fig. 1 (**A**) Cytokine and chemokine profile of monocytes stimulated in vitro with *Salmonella* Enteritidis. CTRL, HIV-NT and HIV-T groups were categorically classified between high and low producers of a given cytokine/chemokine. The diagrams were designed using the value of the global median of each soluble mediator as the cutoff point to mark each individual as "low" or "high" regarding their expression levels of a given analyte. The *black and white diagrams* represent high and low producers of each cytokine, respectively. Each column represents a cytokine and each row represents the pattern of cytokine produced by each patient. Below each table is shown the frequency (%) of individuals with high levels of mediators while the relevant data (>50%) are in *bold and underlined format*. *CTRL* Control, *HIV-NT* HIV non-treated patients, *HIV-T* HIV patients in regular antiretroviral therapy. (**B**) Radar charts summarizing the percentage of high producers for each quantified molecule in the clinical groups studied. The frequency of high producers was greater than 50% (on a scale of 0–100%) in CTRL and HIV-NT groups. *CTRL* Control, *HIV-NT* HIV non-treated, *HIV-T* HIV in regular antiretroviral therapy. The *inner circle* represent 50% of production and the *outer circle* represent the maximum production (100%) of a given analyte

stimulated in vitro with *S*. Enteritidis. From these data, we were able to analyze the cytokine/chemokine profile as described in the **Casuistry and methods** section. The number of subjects with high cytokine levels was compiled in a black-and-white scale diagram to determine the frequency of high producers, in each clinical group. Our data show that CTRL and HIV-NT groups are basically formed by high producers. On the other hand, HIV-T showed a different profile characterized by lower producers for almost all quantified molecules (Fig. 1A).

By using radar charts as a comparison strategy among clinical groups, we observed significant differences in the cytokines/chemokines profile. We considered a relevant cytokine/chemokine production when the frequency of high producers was greater than 50% (Guimarães et al. 2015; Luiza-Silva et al. 2011). Therefore, we observed that after bacteria stimulation, monocytes from CTRL group have more than 50% of high producers for all the cytokines/chemokines evaluated. At the same way, monocytes from HIV-NT group presented a similar profile with

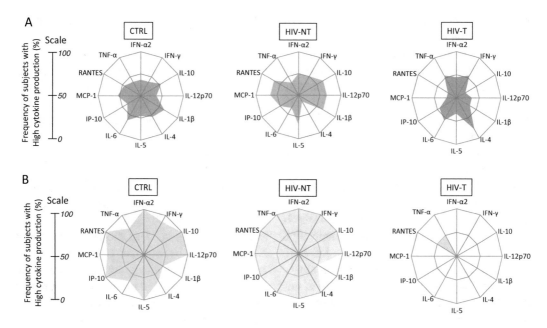

Fig. 2 Radar charts summarizing the percentage of high producers for each quantified molecule in the clinical groups studied shared in (**A**) men and (**B**) women. *CTRL* Control, *HIV-NT* HIV non-treated, *HIV-T* HIV in regular antiretroviral therapy. The *inside circle* represents 50% of production and the *outside circle* represents the maximum production (100%) of a given analyte

more than 50% of high producers for almost all cytokines/chemokines. On the other hand, monocytes from HIV-T patients showed decrease cytokine production, since the majority of individuals in this group were classified as low producers for all the analyzed products (Fig. 1b).

Furthermore, men and women presented a different profile in cytokines production, once that women from CTRL and HIV-NT group produced more cytokines/chemokines than men from the same groups, however, HIV-T women lost the ability to produce cytokines as seen by the lower cytokines/chemokines production compared to HIV-T men (Fig. 2).

6 Discussion

Mononuclear phagocytes contribute to the pathogenesis of HIV (Espíndola et al. 2016); and monocytes/macrophages are one of the two primary cellular targets of HIV-1 (Gartner et al. 1986). These cells support virus replication (Gorry et al. 2005), and act as quiescent and long-lived viral reservoirs (McElrath et al. 1989; Orenstein et al. 1997). Macrophages-derived-monocytes, monocytes from peripheral blood and alveolar macrophages, when infected by HIV-1 show deficiency in effector functions, such as phagocytosis, microbicidal activity and cytokine production (Delemarre et al. 1995; Kedzierska et al. 2001).

HIV-1 infection increase toll like receptor (TLR) expression as well as the signaling in the blood that may progressively affect the immune response and control processes that normally protect individuals from opportunistic infection and disease (Lester et al. 2008). In our results, we observed lower phagocytic function by monocytes from HIV-NT patients compared to CTRL and HIV-T groups, as well as lower microbicidal activity compared to HIV-T patients. Here we are reinforcing that HIV1 infection affects monocytes functions, leading to a biological unbalanced. This was evidenced by the sustained ability to produce high levels of cytokines under bacterial infection and at the same time the inability of developing one of their main effector functions, which is phagocytose and kill the bacteria.

HIV has been related to the modulation of cytokine expression by various cell types, and these cytokines may act in viral replication with stimulatory, inhibitory function or both (Kedzierska et al. 2003).

The breakdown of the gastrointestinal mucosal barrier with translocation of microbial subproducts in the systemic circulation, mainly lipopolysaccharide (LPS), that begins early in the acute phase of HIV infection leading to a progressive rise in the chronic phase of disease is associated with systemic immune activation (Brenchley et al. 2007; Sodora and Silvestri, 2010). LPS interact with CD14/TLR-4 on monocyte/macrophages triggering the secretion of soluble CD14 (sCD14) and pro-inflammatory cytokines such as TNF-α, IL-6 and IL-1 (Kitchens and Thompson 2005).

Furthermore, high levels of viral replication in the gut-associated lymphoid tissue (GALT), is marked by decreased expression of genes regulating epithelial barrier maintenance with consequent increase transcription of immune activation, inflammation, and apoptosis-associated genes, indicating that HIV-induced pathogenesis in GALT contribute for disease progression by impairing the ability to control viral replication and to repair intestinal mucosal tissues (Sankaran et al. 2008).

We had shown high levels of plasmatic cytokines in HIV-1 patients despite antiretroviral therapy (Espíndola et al. 2015). Therefore, we hypothesized that the high levels of systemic cytokines may affect the circulating monocytes functions preventing them from exercising their function properly. Here we showed that HIV-NT patients present similar pattern of cytokines/chemokines production by monocytes compared to CTRL group after stimulation with *Salmonella* Enteritidis. Although, monocytes functions such as phagocytosis and microbicidal activity were impaired during HIV-1 infection in non-treated patients. Interestingly, monocytes from HIV-T patients showed phagocytic and microbicidal capacity partially restored when compared to HIV-NT patients. However, these monocytes

(HIV-T) displayed lower production of cytokines compared to the CTRL group, suggesting that the antiretroviral treatment associated to the reduction of viral load may modulate the immune system in an attempt to reduce chronic inflammation already described in HIV-infected patients.

Besides that, we also observed differences on monocytes functions from HIV1-infected men and women, beyond the difference in the cytokines/chemokines production. It has been shown that male and female hormones may influence the immune response in different situations (Markle et al. 2013; Rodriguez-Garcia et al. 2013; Wira et al. 2011). Here we reported that HIV-1 infected women presented lower phagocytic and microbicidal activity than men (Table 2). Beyond that, women presented increased cytokines/chemokines production than men, however the opposite was observed after treatment, once the cytokines/chemokines production by woman was lower than cytokines/chemokines production by men (Fig. 2). It has been suggested that progesterone, a female hormone, should participate in HIV-1 infection by diminishing the infectivity of HIV and the expression of CCR5 and CXCR4 in cell membrane (Cabrera-Muñoz et al. 2012; Vassiliadou et al. 1999). It has been showed that immune activation represents a strong predictor for HIV-1 disease progression, independent of viral load. Plasmacytoid dendritic cells (pDCs) can contribute importantly to HIV-1–associated immune activation via TLR7 (Beignon et al. 2005; Heil et al. 2004). It has been demonstrated substantial gender-based differences in the response of pDCs to HIV-1 once pDCs derived from women produced significantly higher amounts of IFN-α in response to HIV-1–derived TLR7/8 ligands compared with men (Meier et al. 2009). Furthermore, evidence links immune activation in HIV-1 infection to non–AIDS-related co-morbidities, cardiovascular events, and premature aging. Recent studies indicated that relative increases in heart attack rates are higher among HIV-1–infected women relative to non-infected women, compared with HIV-1–infected men relative to non-infected men (Lang et al. 2010; Triant et al. 2007).

7   Conclusion

Monocytes from HIV non-treated patients showed functional unbalance, since in the presence of a microorganism related to opportunistic infection, they were unable to properly phagocyte and kill the bacteria. Whilst, these monocytes were high producers for cytokines and chemokines such as IL-12, IP-10, MCP-1, RANTES, IL-10 and IL-5, contributing to the cytokine storm characteristic of HIV infection. However, there remain large gaps in our understanding of the precise mechanisms leading to sex based differences in immunity of HIV-1 as well as susceptibility and outcome, and how these can be therapeutically modulated. A better understanding of these mechanisms will be critical to consider gender specific factors in clinical studies involving HIV-infected populations.

Acknowledgements This work was supported by the São Paulo Research Foundation (FAPESP #2011/12199-0), the National Council for Scientific and Technological Development (CNPq), and Coordination for the Improvement of Higher Education Personnel (Capes). The authors are grateful to Dr. Elyara Maria Soares for her critical reading of the manuscript.

Competing Interests The authors have no competing interests.

Author Contribution Statement MCC designed and performed experiments and wrote the paper; MSE, LJGL, LSS, CF contributed by technical assistance, standardization of techniques and advices and discussion of data; MAP assisted in the statistical analises; FC and JPF provided the bacterial strain and developed all techniques of bacterial culture and isolation; VRB is the physician responsible for selecting, consulting patients, and analyzing clinical data; FGF developed the concept, designed the experiments, supervised the study, and wrote the paper.

References

Barré-Sinoussi F, Chermann JC, Rey F, Nugeyre MT, Chamaret S, Gruest J, Dauguet C, Axler-Blin C, Vézinet-Brun F, Rouzioux C et al (1983) Isolation of a T-lymphotropic retrovirus from a patient at risk for acquired immune deficiency syndrome (AIDS). Science 220:868–871

Beignon A-S, McKenna K, Skoberne M, Manches O, DaSilva I, Kavanagh DG, Larsson M, Gorelick RJ, Lifson JD, Bhardwaj N (2005) Endocytosis of HIV-1 activates plasmacytoid dendritic cells via Toll-like receptor-viral RNA interactions. J Clin Invest 115:3265–3275

Bhagani S (2011) Antimicrobial therapy for the treatment of opportunistic infections in HIV/AIDS patients: a critical appraisal. HIVAIDS – Res Palliat Care 19

Brenchley JM, Price DA, Schacker TW, Asher TE, Silvestri G, Rao S, Kazzaz Z, Bornstein E, Lambotte O, Altmann D et al (2007) Microbial translocation is a cause of systemic immune activation in chronic HIV infection. Nat Med 12:1365–1371

Burdo TH, Lentz MR, Autissier P, Krishnan A, Halpern E, Letendre S, Rosenberg ES, Ellis RJ, Williams KC (2011) Soluble CD163 made by monocyte/macrophages is a novel marker of HIV activity in early and chronic infection prior to and after antiretroviral therapy. J Infect Dis 204:154–163

Cabrera-Muñoz E, Fuentes-Romero LL, Zamora-Chávez J, Camacho-Arroyo I, Soto-Ramírez LE (2012) Effects of progesterone on the content of CCR5 and CXCR4 coreceptors in PBMCs of seropositive and exposed but uninfected Mexican women to HIV-1. J Steroid Biochem Mol Biol 132:66–72

Campioni F, Moratto Bergamini AM, Falcão JP (2012) Genetic diversity, virulence genes and antimicrobial resistance of Salmonella Enteritidis isolated from food and humans over a 24-year period in Brazil. Food Microbiol 32:254–264

Clumeck N, Mascart-Lemone F, de Maubeuge J, Brenez D, Marcelis L (1983) Acquired immune deficiency syndrome in Black Africans. Lancet Lond Engl 1:642

Delemarre FG, Stevenhagen A, Van Furth R (1995) Granulocyte-macrophage colony-stimulating factor (GM-CSF) reduces toxoplasmastatic activity of human monocytes via induction of prostaglandin E2 (PGE2). Clin Exp Immunol 102:425–429

Dhanoa A, Fatt QK (2009) Non-typhoidal Salmonella bacteraemia: epidemiology, clinical characteristics and its' association with severe immunosuppression. Ann Clin Microbiol Antimicrob 8:15

Espíndola MS, Lima LJG, Soares LS, Cacemiro MC, Zambuzi FA, de Souza Gomes M, Amaral LR, Bollela VR, Martins-Filho OA, Frantz FG (2015) Dysregulated immune activation in second-line HAART HIV+ patients is similar to that of untreated patients. PLoS One 10:e0145261

Espíndola MS, Soares LS, Galvão-Lima LJ, Zambuzi FA, Cacemiro MC, Brauer VS, Frantz FG (2016) HIV infection: focus on the innate immune cells. Immunol Res 64:1118

Farzadegan H, Hoover DR, Astemborski J, Lyles CM, Margolick JB, Markham RB, Quinn TC, Vlahov D (1998) Sex differences in HIV-1 viral load and progression to AIDS. Lancet Lond Engl 352:1510–1514

Gartner S, Markovits P, Markovitz DM, Betts RF, Popovic M (1986) Virus isolation from and identification of HTLV-III/LAV-producing cells in brain tissue from a patient with AIDS. JAMA 256:2365–2371

Gorry PR, Sonza S, Kedzierska K, Crowe SM (2005) Isolation of human immunodeficiency virus type 1 from peripheral blood monocytes. Methods Mol Biol Clifton NJ 304:25–33

Guimarães AGDP, da Costa AG, Martins-Filho OA, Pimentel JPD, Zauli DAG, Peruhype-Magalhães V, Teixeira-Carvalho A, Béla SR, Xavier MAP, Coelho-Dos-Reis JG et al (2015) CD11c+CD123Low dendritic cell subset and the triad TNF-α/IL-17A/IFN-γ integrate mucosal and peripheral cellular responses in HIV patients with high-grade anal intraepithelial neoplasia: a systems biology approach. J Acquir Immune Defic Syndr 1999(68):112–122

Heil F, Hemmi H, Hochrein H, Ampenberger F, Kirschning C, Akira S, Lipford G, Wagner H, Bauer S (2004) Species-specific recognition of single-stranded RNA via toll-like receptor 7 and 8. Science 303:1526–1529

Kedzierska K, Mak J, Jaworowski A, Greenway A, Violo A, Chan HT, Hocking J, Purcell D, Sullivan JS, Mills J et al (2001) nef-deleted HIV-1 inhibits phagocytosis by monocyte-derived macrophages in vitro but not by peripheral blood monocytes in vivo. AIDS Lond Engl 15:945–955

Kedzierska K, Crowe SM, Turville S, Cunningham AL (2003) The influence of cytokines, chemokines and their receptors on HIV-1 replication in monocytes and macrophages. Rev Med Virol 13:39–56

Kitchens RL, Thompson PA (2005) Modulatory effects of sCD14 and LBP on LPS-host cell interactions. J Endotoxin Res 11:225–229

Klein SL (2012) Sex influences immune responses to viruses, and efficacy of prophylaxis and treatments for viral diseases. BioEssays News Rev Mol Cell Dev Biol 34:1050–1059

Kuroda MJ (2010) Macrophages: do they impact AIDS progression more than CD4 T cells? J Leukoc Biol 87:569–573

Lang S, Mary-Krause M, Cotte L, Gilquin J, Partisani M, Simon A, Boccara F, Bingham A, Costagliola D, French Hospital Database on HIV-ANRS CO4 (2010) Increased risk of myocardial infarction in HIV-infected patients in France, relative to the general population. AIDS Lond Engl 24:1228–1230

Lester RT, Yao X-D, Ball TB, McKinnon LR, Kaul R, Wachihi C, Jaoko W, Plummer FA, Rosenthal KL (2008) Toll-like receptor expression and responsiveness are increased in viraemic HIV-1 infection. AIDS Lond Engl 22:685–694

Luiza-Silva M, Campi-Azevedo AC, Batista MA, Martins MA, Avelar RS, da Silveira Lemos D, Bastos Camacho LA, de Menezes Martins R, de Lourdes de Sousa Maia M, Guedes Farias RH et al (2011) Cytokine signatures of innate and adaptive immunity in 17DD yellow fever vaccinated children and its association with the level of neutralizing antibody. J Infect Dis 204:873–883

MacLennan CA, Gilchrist JJ, Gordon MA, Cunningham AF, Cobbold M, Goodall M, Kingsley RA, van Oosterhout JJG, Msefula CL, Mandala WL et al (2010) Dysregulated humoral immunity to nontyphoidal Salmonella in HIV-infected African adults. Science 328:508–512

Markle JGM, Frank DN, Mortin-Toth S, Robertson CE, Feazel LM, Rolle-Kampczyk U, von Bergen M, McCoy KD, Macpherson AJ, Danska JS (2013) Sex differences in the gut microbiome drive hormone-dependent regulation of autoimmunity. Science 339:1084–1088

McElrath MJ, Pruett JE, Cohn ZA (1989) Mononuclear phagocytes of blood and bone marrow: comparative roles as viral reservoirs in human immunodeficiency virus type 1 infections. Proc Natl Acad Sci U S A 86:675–679

Meditz AL, MaWhinney S, Allshouse A, Feser W, Markowitz M, Little S, Hecht R, Daar ES, Collier AC, Margolick J et al (2011) Sex, race, and geographic region influence clinical outcomes following primary HIV-1 infection. J Infect Dis 203:442–451

Meier A, Chang JJ, Chan ES, Pollard RB, Sidhu HK, Kulkarni S, Wen TF, Lindsay RJ, Orellana L, Mildvan D et al (2009) Sex differences in the Toll-like receptor-mediated response of plasmacytoid dendritic cells to HIV-1. Nat Med 15:955–959

Meléndez LM, Colon K, Rivera L, Rodriguez-Franco E, Toro-Nieves D (2011) Proteomic analysis of HIV-infected macrophages. J Neuroimmune Pharmacol Off J Soc NeuroImmune Pharmacol 6:89–106

Orenstein JM, Fox C, Wahl SM (1997) Macrophages as a source of HIV during opportunistic infections. Science 276:1857–1861

Preziosi MJ, Kandel SM, Guiney DG, Browne SH (2012) Microbiological analysis of nontyphoidal Salmonella strains causing distinct syndromes of bacteremia or enteritis in HIV/AIDS patients in San Diego. California J Clin Microbiol 50:3598–3603

Rodriguez-Garcia M, Patel MV, Wira CR (2013) Innate and adaptive anti-HIV immune responses in the female reproductive tract. J Reprod Immunol 97:74–84

Sankaran S, George MD, Reay E, Guadalupe M, Flamm J, Prindiville T, Dandekar S (2008) Rapid onset of intestinal epithelial barrier dysfunction in primary human immunodeficiency virus infection is driven by an imbalance between immune response and mucosal repair and regeneration. J Virol 82:538–545

Sodora DL, Silvestri G (2010) HIV, mucosal tissues, and T helper 17 cells: where we come from, where we are, and where we go from here. Curr Opin HIV AIDS 5:111–113

Sterling TR, Lyles CM, Vlahov D, Astemborski J, Margolick JB, Quinn TC (1999) Sex differences in longitudinal human immunodeficiency virus type 1 RNA levels among seroconverters. J Infect Dis 180:666–672

Thompson MA, Aberg JA, Hoy JF, Telenti A, Benson C, Cahn P, Eron JJ, Günthard HF, Hammer SM, Reiss P et al (2012) Antiretroviral treatment of adult HIV infection: 2012 recommendations of the International Antiviral Society-USA panel. JAMA 308:387–402

Touloumi G, Pantazis N, Babiker AG, Walker SA, Katsarou O, Karafoulidou A, Hatzakis A, Porter K, CASCADE Collaboration (2004) Differences in HIV RNA levels before the initiation of antiretroviral therapy among 1864 individuals with known HIV-1 seroconversion dates. AIDS Lond Engl 18:1697–1705

Triant VA, Lee H, Hadigan C, Grinspoon SK (2007) Increased acute myocardial infarction rates and cardiovascular risk factors among patients with human immunodeficiency virus disease. J Clin Endocrinol Metab 92:2506–2512

UNAIDS. *Global AIDS Update 2016.*

Vassiliadou N, Tucker L, Anderson DJ (1999) Progesterone-induced inhibition of chemokine receptor expression on peripheral blood mononuclear cells correlates with reduced HIV-1 infectability in vitro. J Immunol Baltim Md 1950(162):7510–7518

Wira CR, Ghosh M, Smith JM, Shen L, Connor RI, Sundstrom P, Frechette GM, Hill ET, Fahey JV (2011) Epithelial cell secretions from the human female reproductive tract inhibit sexually transmitted pathogens and Candida albicans but not Lactobacillus. Mucosal Immunol 4:335–342

Zybarth G, Reiling N, Schmidtmayerova H, Sherry B, Bukrinsky M (1999) Activation-induced resistance of human macrophages to HIV-1 infection in vitro. J Immunol Baltim Md 1950(162):400–406

Adv Exp Med Biol - Advances in Microbiology, Infectious Diseases and Public Health (2018) 9: 41–45
DOI 10.1007/5584_2017_47
© Springer International Publishing AG 2017
Published online: 1 June 2017

A Synonymous Mutation at Bovine Alpha Vitronectin Domain of Integrin Host Receptor (ITGAV) Gene Effect the Susceptibility of Foot-and-Mouth Disease in Crossbred Cattle

Rani Singh, Rani Alex, Umesh Singh, Sushil Kumar, Gyanendra Singh Sengar, T.V. Raja, R.R. Alyethodi, Ashish Kumar, and Rajib Deb

Abstract

Integrins are one of the major biologically active proteins responsible for Foot-and-mouth disease virus (FMDV)- host interaction. Out of various heterodimeric integrins discovered, αVβ6 heterodimer serves as the chief receptor for FMDV host tropism. Earlier studies reported that, SNPs at beta 6 subunit (ITGB6) were associated with the occurrence of the diseases in cattle. In this study we report the association between a synonymous SNP (rs719257875) at bovine alpha vitronectin domain of integrin receptor (ITGAV) gene and FMD susceptibility in cattle. A strong significant association ($P < 0.0001$) of the genotypes with FMD susceptibility were obtained, where the CC genotypes play a major role in occurrence of FMD in crossbred cattle.

Keywords

ITGAV · FMDV · Synonymous SNP · Cattle

R. Singh, R. Alex, U. Singh, S. Kumar, G.S. Sengar,
T.V. Raja, R.R. Alyethodi, A. Kumar, and R. Deb (✉)
Molecular Genetics Laboratory,
ICAR-Central Institute for Research on Cattle,
Meerut 250 001, Uttar Pradesh, India
e-mail: drrajibdeb@gmail.com

Foot-and-mouth disease (FMD) is a highly contagious and economically annihilating disease of cloven-hoofed livestock. The preventive measures to control the incidence of this disease often fail because the causative FMD virus (FMDV) mutates recurrently evolving new subtypes (Anonymous 2008). The control of the disease in the Asian region via inactivated vaccine leads into unsatisfactory results due to various reasons (Rodriguez and Gay 2011) and technical difficulties (Seneque 2011). Further, in the recent past FMD has made its appearance even in the FMD free countries (Knowles et al. 2012) which hints at the depth of the devastating nature of this disease and urges for alternate approaches for disease resistance. A novel and recent approach is development of FMD resistant cattle, is imminent by using marker-assisted selection (MAS) after identifying suitable candidate genes related to disease resistant. The role and importance of integrins for initiation of FMD infection is well studied (O'donnell et al. 2009).

Integrins are a family of heterodimeric integral membrane proteins composed of an α and a β subunit and functions in cell surface adhesion and signaling. The proteolytically generated αV subunit associates with β1, β3, β5, β6 and β8 subunits to form heterodimeric αVβ1, αVβ2, αVβ3, αVβ6, and αVβ8 integrins receptors via a highly conserved Arginine-glycine-aspartic acid (RGD) amino acid sequence motif (O'donnell et al. 2009). Of the various heterodimeric integrins, αVβ6 heterodimer serves as the major receptor that determines the tropism of FMDV for the epitheliall cells (O'donnell et al. 2009).

The Zebu (*Bos indicus*) cattle are less susceptible to FMD than taurine (*Bos taurus*) (Longjam et al. 2011). Yet, the genetic basis which contributes to the FMD resistant phenotype for the bovine host is not studied effusively. In our earlier study, we identified the effect of two SNPs (rs13500299 and rs109075046) located at beta 6 domain of bovine Integrin (ITGB6) receptor gene on FMDV susceptibility among crossbreed cattle (Singh et al. 2014, 2015a, b) . Based on the evidence described previously, the present work aimed to identify the effect of genetic variation exist within the alpha vitronectin domain of bovine Integrin (ITGAV) receptor gene on the occurrence of FMD among crossbreed cattle. Bovine ITGAV is located on chromosome 2 and consists of thirty exons with a size of 3147 bp. On analysis of the entire bovine ITGAV gene, we pinpoint a synonymous mutation at 26th exon (rs719257875) is associated with Foot and Mouth Disease Virus (FMDV) susceptibility in crossbreed cattle.

Blood samples were collected in the jugular vein of different indigenous, viz. Sahiwal (n = 51), Kankrej (n = 48), Ongole (n = 28) and crossbreed (n = 142) cattle. In an outbreak of FMD, a total 67 blood samples were collected from crossbreed bulls from the Bull Rearing Unit, Central Institute for Research on Cattle, Meerut, India. The animals are divided into two groups base on the FMD incidence. The first group consisted 35 animals (case n = 35) that are infected by FMD and the second group having 31 animals (control n = 31) that are non-infected with FMD. A r3AB NSP ELISA test was used for the confirmation of all FMD infected samples. After collection, the blood samples were stored at −80 °C until DNA extraction. The genomic DNA was extracted by using phenol-chloroform extraction method (Sambrook and Russell 2001) and the purity was checked by spectrophotometrically.

In the present study we targeted a synonymous SNP (rs719257875: GAC to GAT, both codes for Asparatic acid) at 26th exonic region of *Bos taurus* ITGAV through ensemble genome browser analysis. The genotype of the targeted synonymous SNP (rs719257875) was screened by using PCR RFLP procedure. Primers were designed according to *Bos, tauru* ITGAV sequence (http://asia.ensembl.org/index.html). The specificity of the primers was checked through 'BLAST' program (http://www.ncbi.nlm.nih.gov/blast). The primers used in this study were as follows: ITGAV 26 F: TTCTGTGCTTAGTGTTAGCC and ITGAV 26R: AGAGGAAAGTAAGAGGAGTG. The PCR was performed in a 25 μl reaction mixture containing 1X PCR buffer (Sigma aldrich, USA), 1.5 mM MgCl₂ (Sigma Aldrich, USA), 200 μM dNTPs (Sigma aldrich, USA), 10 pmol of each primers, 1 U Taq DNA polymerase (Sigma aldrich, USA) and 50 ng of genomic DNA as template. The cycling protocol was initial denaturation for 5 min at 95 °C followed by 35 cycles (94 °C for 50 s, 53 °C annealing for 35 s, 72 °C for 35 s), with a final extension at 72 °C for 10 min. The amplified products were run along with 100 bp DNA ladder on 1.5% agarose gel, stained with ethidium bromide and the bands were visualized under UV (AlphaImager® Gel Documentation) for documentation. For genotyping, PCR product was digested with HpyCH4IVrestriction enzyme to determine ITGAV alleles at 26th exon. Gene fragments were subjected to digestion by restriction enzymes in a total volume of 20 μl (8 μl PCR products, 1 X enzyme buffers, 1 U enzymes and distilled water) and placed in the incubator at 37 °C for overnight. The digested PCR products were analyze by 3% agarose gel electrophoresis. For Further validation the PCR products of each genotype were purified by PCR purification kit

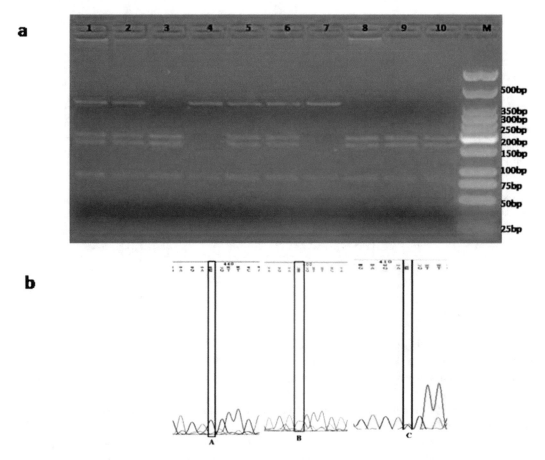

Fig. 1 PCR- HpyCH4IV RFLP based genotypic varia-tion of SNP (rs719257875) at alpha vitronectin domain of Integrin (ITGAV) host receptor gene. (**a**) Agarose gel electrophoresis of the digested PCR product. Lane1, 2, 5 & 6: CT (82, 159, 180 & 339 bp); Lane 3, 8, 9&10: CC (82,159& 180 bp); Lane 4&7: TT (82 & 339 bp). (**b**) Dendrogram analysis for the different genotypes sequenced. *A*, CC (KU951575); *B*, TT (KU951576); and *C*, CT (KU951577)

(Qiagen, Germany) and subjected to TA cloning using pTZ57R plasmid (InsTA clone PCR Clon-ing Kit, Fermentas). Each clone was confirmed by colony PCR, and DNA sequencing (Xcelris Labs Ltd., India). For each breed allele and geno-type frequencies were calculated based on direct counting. Further, the differences between geno-type of the breeds were performed by Chi-square test. Fischer's exact test was used to analyze the effect of genotype on the incidence of FMD.

HpyCH4IV digestion of amplified products showed three fragments (180, 159 and 82 bp) for CC genotype, two for TT (339 and 82) and four fragments (339, 180, 158 and 82 bp) for CT genotype (Fig. 1). Further confirmation of the

respective genotypes was assessed through DNA sequencing and results were submitted to NCBI Gen Bank (CC, KU951575; TT, KU951576; CT, KU951577). The overall genotypes and allele frequencies in different breeds were presented in Table 1. Chi-square analysis indicated that the genotype frequencies varied between different breeds. In order to ensure the difference in genotypic frequencies among indigenous and crossbred cattle, the geno-typic frequencies from different native breeds were merged and compared with the crossbred population by chi-square tests. The genotype frequencies of CC, TT, and CT in the indigenous and crossbreed population were 0.326, 0.605 and

Table 1 Genotype and allele frequency of the synonymous mutation at 26th Exon of the ITGAV gene among different among cattle breeds

Breed	Allele frequencies		Genotype frequency			Total	Chi-Square
	C	T	CC	CT	TT		32.66**
Kankrej	0.552	0.448	10(0.208)	33(0.687)	5(0.104)	48	
Ongole	0.535	0.465	6(0.214)	18(0.642)	4(0.142)	28	
Sahiwal	0.662	0.338	28(0.394)	38(0.535)	5(0.07)	71	
Crossbreed	0.782	0.218	83(0.584)	56(0.394)	3(0.021)	142	
Indigenous	0.602	0.398	44(0.326)	89(0.605)	14(0.095)	147	17.17**

**P < 0.00001; significant

Table 2 Genotype and allele frequency of the synonymous mutation at 26th exon of the ITGAV gene among FMDV-infected (case) and non-infected (control)

Status	Allele frequence		Genotype frequency			Total	Fischer's exact test(P value)
	C	T	CC	CT	TT		0.00043**
Case	0.914	0.086	29(0.828)	6(0.171)	0(0)	35	
Control	0.677	0.323	12(0.387)	18(0.580)	1(0.032)	31	
Total			41	24	1		

**P < 0.0001; significant

0.095 and 0.584, 0.394 and 0.021, respectively. The H-W equilibrium analysis showed that crossbreed population was in H-W equilibrium while the indigenous populations showed a deviations from H-W equilibrium which may be due to natural selection acting on them for disease resistance while the crossbreed are mostly selected for production performance. The genotypic frequencies varied between indigenous and crossbred population. The frequency of T allele was higher in indigenous (0.398) than crossbreed (0.218) population under study.

The genotype frequencies of ITGAV gene among FMD-infected (case) animals and non-infected (control) were estimated and presented in Table 2. Both groups were in H-W equilibrium. Out of 35 FMD infected animals, the frequencies for the CC, CT and TT genotypes were 0.828, 0.171 and 0 respectively. In the non-infected animals, the estimated genotypic frequencies of the three genotypes in the same order were 0.387, 0.580 and 0.032. The Fischer's exact analysis revealed a strong association between the genotype and incidences of disease. The frequency of CC genotype was significantly higher in infected animals while the T allele appeared only in the heterozygous pattern in infected animals.

Historically, silent mutations were thought to be of little to no significance. However, certain scientific reports suggests synonymous mutations could change protein translation efficiency and protein folding as well as its function (Czech et al. 2010), mRNA stability (Angov 2011). Genome-wide association studies have revealed a substantial contribution of synonymous SNPs to human disease risk and other complex traits (Zuben and Chava 2011). Similar way, using the synonymous mutations, viral attenuation has been achieved (Coleman et al. 2008). The present study gives insights on the association of the SNP (rs719257875) on the FMD susceptibility. Strong association (P < 0.0001) of the genotypes with FMD susceptibility, suggests that the CC genotypic groups are more susceptible to FMD compare to its counterparts. Present findings may thus provide some reference to be implemented for FMD disease resistant breeding.

Acknowledgement This project was funded by Department of Science and Technology, New Delhi, Govt. of India (SR/WOS-A/LS-437/2012 (G)). The authors are thankful to Director, ICAR-CIRC, Meerut for providing necessary facilities for conducting the study.

Ethical Standard Statement All animal experiments were performed following protocols approved by the Institutional Ethical committee.

Conflict of Interest We declared that none of the authors have any kind of conflict of interest interest for submission of this present manuscript.

References

Angov E (2011) Codon usage: nature's roadmap to expression and folding of proteins. Biotechnol J 6 (6):650–659

Anonymous (2008) Foot-and-mouth disease. In: Manual of diagnostic tests and vaccines for terrestrial animals (mammals, birds and bees). World Organization for Animal Health (OIE), Paris

Coleman JR, Papamichail D, Skiena S, Futcher B, Wimmer E, Mueller S (2008) Virus attenuation by genome-scale changes in codon pair bias. Science 27,320(5884):1784–1787. doi:10.1126/science. 1155761

Czech A, Fedyunin I, Zhang G, Ignatova Z (2010) Silent mutations in sight: co-variations in tRNA abundance as a key to unravel consequences of silent mutations. Mol BioSyst 6(10):1767–1772. doi:10.1039/c004796c

Knowles NJ, He J, Shang Y, Wadsworth J, Valdazo-Gonzalez B, Onosato H et al (2012) Southeast Asian foot-and-mouth disease viruses in Eastern Asia. Emerg Infect Dis 18:499–501

Longjam N, Deb R, Sarmah AK, Tayo T, Awachat VB, Saxena VK (2011) A brief review on diagnosis of foot-and-mouth disease of livestock: conventional to molecular tools. Vet Med Int 2011(1-2):1–17

O'donnell V, Pacheco JM, Gregg D, Baxt B (2009) Analysis of foot-and-mouth disease virus integrin receptor expression in tissues from naive and infected cattle. J Comp Pathol 141(2):98–112

Rodriguez LL, Gay CG (2011) Development of vaccines toward the global control and eradication of foot-and-mouth disease. Expert Rev Vaccines 10:377–387

Sambrook J, Russell DW (2001) Molecular cloning: a laboratory manual. Cold Spring Harbor, New York

Seneque S (2011) Foot-and-Mouth Disease control in Asia. Meeting unique challenges. In: Proceedings of the proceedings of 5th Asian Pig veterinary society congress Pattaya, Thailand

Singh R, Deb R, Singh U, Alex R, Kumar S, Chakraborti S, Sharma S, Sengar G, Singh R (2014) Development of a tetra-primer ARMS PCR-based assay for detection of a novel single-nucleotide polymorphism in the 5′ untranslated region of the bovine ITGB6 receptor gene associated with foot-and-mouth disease susceptibility in cattle. Arch Virol 159 (12):3385–3389. doi:10.1007/s00705-014-2194-0

Singh R, Deb R, Singh U, Raja TV, Alex R, Kumar S, Chakraborti S, Alyethodi RR, Sharma S, Sengar G (2015a) Heterozygosity at the SNP (rs136500299) of ITGB6 receptor gene possibly influences the susceptibility among crossbred bull to foot and mouth disease infection. Virus Dis 26(1–2):48–54. doi:10.1007/s13337-015-0249-9

Singh R, Deb R, Singh U, Raja TV, Alex R, Kumar S, Chakraborti S, Sengar G, Sharma S (2015b) Single tube tetraplex PCR based screening of a SNP at exon 14 region of bovine ITGB6 among different Zebu breeds. Meta Gene. 10(3):26–30. doi:10.1016/j.mgene.2014.12.004

Zuben ES, Chava K-S (2011) Understanding the contribution of synonymous mutations to human disease. Nat Rev Genet 12:683–691

Adv Exp Med Biol - Advances in Microbiology, Infectious Diseases and Public Health (2018) 9: 47–62
DOI 10.1007/5584_2017_53
© Springer International Publishing AG 2017
Published online: 11 June 2017

Antibiotic Resistance Genes and Antibiotic Susceptibility of Oral *Enterococcus faecalis* Isolates Compared to Isolates from Hospitalized Patients and Food

Annette Carola Anderson, Huria Andisha, Elmar Hellwig, Daniel Jonas, Kirstin Vach, and Ali Al-Ahmad

Abstract

Enterococcus faecalis, a commensal of the intestinal tract of humans and animals is of great significance as leading opportunistic pathogen, and also prevalent in oral diseases, such as endodontic infections, as well as the healthy oral cavity. To investigate the potential of oral E. faecalis to constitute a reservoir of antibiotic resistance, isolates from supragingival plaque/saliva and from endodontic infections were screened regarding their resistance to selected antibiotics in comparison to nosocomial and food isolates.

70 E. faecalis isolates were analyzed with PCR regarding their equipment with the resistance genes tetM, tetO, ermB, ermC, vanA, vanB and blaTEM. Additionally, they were tested for their phenotypic resistance to doxycycline, azithromycin, rifampicin, amoxicillin and streptomycin using the Etest.

High percentages of the plaque/saliva, nosocomial and food isolates were resistant to doxycycline and azithromycin, particularly plaque/saliva isolates (81%) and nosocomial isolates (73.3%) showed resistance to doxycycline, significantly more than among the food and endodontic isolates. Rifampicin resistance was widespread among isolates from plaque/saliva (52.4%), endodontic infections (50%) and nosocomial infections (40%); all isolates were susceptible to amoxicillin and all oral isolates to high-level streptomycin.

A.C. Anderson (✉), H. Andisha, E. Hellwig,
and A. Al-Ahmad
Department of Operative Dentistry and Periodontology,
Medical Center - University of Freiburg, Faculty of
Medicine, University of Freiburg, Hugstetter Straße 55,
79106 Freiburg, Germany
e-mail: annette.anderson@uniklinik-freiburg.de

D. Jonas
Institute for Environmental Health Sciences and Hospital
Infection Control, Medical Center - University of
Freiburg, Faculty of Medicine, University of Freiburg,
Breisacher Straße 115b, 79106 Freiburg, Germany

K. Vach
Center for Medical Biometry and Medical Informatics,
Institute for Medical Biometry and Statistics, Medical
Center - University of Freiburg, Faculty of Medicine,
University of Freiburg, Stefan-Meier Straße 26, 79104
Freiburg, Germany

TetM genes were detected in the majority of all isolates and ermB genes were present in many nosocomial and plaque/saliva isolates. Thirty percent of the endodontic isolates and 53% of the nosocomial isolates were equipped with blaTEM genes.

The results suggest that the oral cavity can harbor E. faecalis strains with multiple resistances against different antibiotics and thus be regarded as a potential source of resistance traits.

Keywords

Antibiotic resistance · Endodontic infections · Enterococci · Food · Nosocomial infections · Oral cavity · Plaque

1   Introduction

Enterococcus faecalis is of great importance as a leading opportunistic pathogen causing nosocomial infections (Arias and Murray 2012). Frequent infections include endocarditis, meningitis, urinary tract infections, wound infections and neonatal infections (Murray 1990). Furthermore, this bacterial species is of significance for the field of oral diseases. Even though *E. faecalis* is normally only found in low numbers in oral sites of healthy individuals (Portenier et al. 2003), its prevalence in the oral cavity substantially increases in many oral diseases, e.g. gingivitis, periodontitis, caries, endodontic infections and especially post-treatment apical periodontitis, where it is considered a main pathogen associated with endodontic failure (Dahlen et al. 2000; C. M. Sedgley et al. 2005b; C. Sedgley et al. 2006; Souto and Colombo 2008; Sun et al. 2009; Kouidhi et al. 2011; Anderson et al. 2013).

What contributes to its relevance as an opportunistic pathogen is the fact that *E. faecalis* is intrinsically resistant to several antibiotics and can harbor different acquired resistance traits (Van Tyne and Gilmore 2014). Despite its pathogenic potential, *E. faecalis* is typically found as a commensal in the gastrointestinal tract of humans and many animals (Arias and Murray 2012). Specific strains have been used as probiotics and some strains are found in food where they are responsible e.g. for the ripening of certain cheeses (Fisher and Phillips 2009; Franz et al. 2011; Hammerum 2012).

Its proficiency in efficiently acquiring and spreading genetic elements via horizontal gene transfer as well as its common ability to form biofilms have been well characterized for *E. faecalis* (Paulsen et al. 2003; Duggan and Sedgley 2007; Manson et al. 2010; Palmer et al. 2010; Paganelli et al. 2012) . Our group was able to demonstrate that *E. faecalis* originating from cheese is able to integrate into the oral biofilm *in vivo* (Al-Ahmad et al. 2010) and recently authors confirmed that *E. faecalis* can colonize a multi-species biofilm in a supragingival biofilm model (Thurnheer and Belibasakis 2015). These findings highlight the possibility of the oral cavity to constitute a reservoir for the antibiotic resistance genes of *E. faecalis* as well as other traits of clinical concern that could be spread within the oral biofilm. The information on antibiotic susceptibility characteristics of oral *E. faecalis* isolates is scarce apart from studies of endodontic isolates. In a previous study we investigated the virulence factors as well as the capacity for biofilm formation and susceptibility to some antibiotics of the *E. faecalis* isolates from different sources with a focus on the biofilm formation in association with virulence factors (Anderson et al. 2015). The aim of the present study was to take this analysis further focusing on selected antibiotic resistance genes as well as additional relevant phenotypic resistance to assess whether these strains can represent a reservoir for antibiotic resistance traits. For the determination of the antibiotic susceptibility, widely used phenotypic tests as well as PCR were applied (Amsler et al. 2010; Jorgensen and Ferraro 2009; Tenover et al. 1996), which give an insight into the bacterial strains' equipment with resistance traits. This is advantageous in the attempt to determine if strains possessed genes that could be spread even without expressing the genes themselves.

2   Materials and Methods

2.1   Bacterial Isolates

A total of 70 *E. faecalis* strains, 20 isolates from endodontic retreatment, 21 oral isolates (saliva and supragingival plaque), 14 food isolates (raw milk) and 15 isolates from nosocomial infections (9 urinary tract infections, 1 wound infection, 1 intraoperative swab, 1 drainage secretion, 1 intraabdominal aspirate, 1 blood culture and 1 central venous catheter) were analyzed for their antibiotic susceptibility. The oral and endodontic isolates were gathered from 2011–2014 in the Department of Operative Dentistry and Periodontology (Medical Center - University of Freiburg, Faculty of Medicine, University of Freiburg, Germany). The plaque and saliva samples were taken from healthy individuals with the following exclusion criteria: no serious illness, no use of antibiotics until 6 months prior to the study, no pregnancy or lactation, healthy oral status and absence of carious lesions. The food isolates from raw milk samples from different cows were received in 2014 from the Bavarian Health and Food Safety Authority (Oberschleißheim, Germany) and the isolates from nosocomial infections were obtained from patients in 2013 from the Department of Medical Microbiology and Hygiene (Medical Center - University of Freiburg, Faculty of Medicine, University of Freiburg, Germany). All endodontic and clinical isolates were obtained after approval by the Ethics Committee (no. 140/09, University of Freiburg) and the collection of the

endodontic samples followed the protocol from Schirrmeister et al. (2007). Prior to the antibiotic susceptibility testing, the isolates were confirmed to be *E. faecalis* by amplification of a species-specific 16S rDNA fragment (Table 1). The following strains were used as reference strains: *Klebsiella pneumoniae* 1230 and *Enterobacter cloacae* 472 (containing the bla_{TEM} gene), *Enterococcus faecium* 401, *E. faecium* 403 and *E. faecium* 643 (containing the *ermB*-gene), *Staphylococcus aureus* 2250, *S. aureus* 2223, *S. aureus* 4331 (containing the *ermC*-gene), *E. faecium* 633 and *E. faecium* 643 (containing the *vanA*-gene), *E. faecalis* V583, *E. faecium* BM4524, *E. faecium* 401 (containing the *vanB*-gene), *E. faecium* 633 (containing the *tetM*-gene) and a bacterial isolate from sewage MG (containing the *tetO*-gene). All reference strains were kindly provided by Prof. Daniel Jonas (Institute for Environmental Health Sciences and Hospital Infection Control, Medical Center - University of Freiburg, Faculty of Medicine, University of Freiburg, Germany).

2.2   Antibiotic Susceptibility Testing with the Etest

The Etest (Liofilchem, Roseto degli Abruzzi, Italy) was used to test the susceptibility of all the *E. faecalis* isolates to the following antibiotics: Doxycycline, azithromycin, amoxicillin, rifampicin and high-level streptomycin. The method was conducted according to the manufacturer's protocol as described earlier (Al-Ahmad et al. 2014). Specifically, material from pure colonies was taken from an overnight culture, suspended in sterile NaCl-solution (0.9%) to reach an inoculum turbidity of McFarland 0.5. Each sample was streaked on Mueller-Hinton agar plates and one Etest strip was placed on each agar plate using sterile tweezers. The results were interpreted as indicating

Table 1 Primers used for the detection of different antibiotic resistance genes of *E. faecalis* by PCR

Target	Primer	Primer sequence (5′–3′)	Amplicon size [bp]	References
E. faecalis	Efaec-F	GTTTATGCCGCATGGCATAAGAG	310	Siqueira and Rocas (2004)
	Efaec-R	CCGTACGGGGACGTTCAG		
blaTEM	blaTEM f	CCAATGCTTAATCAGTGAGG	858	Call et al. (2003)
	blaTEM r	ATGAGTATTCAACATTTCCG		
tetM	tetMf	AGTTTTAGCTCATGTTGATG	1862	Perez-Trallero et al. (2007)
	tetMr	TCCGACTATTTGGACGACGG		
tetO	tetO-f	GCGGAACATTGCATTTGAGGG	538	Perez-Trallero et al. (2007)
	tetO-r	CTCTATGGACAACCCGACAGAAG		
ermB	ermB-f	GAAAAGGTACTCAACCAAATA	639	Reinert et al. (2008)
	ermB-r	AGTAACGGTACTTAAATTGTTTAC		
ermC	ermC-f	AATCGGCTCAGGAAAAGG	562	Perreten et al. (2005)
	ermC-r	ATCGTCAATTCCTGCATG		
vanA	vanA$_1$	GGGAAAACGACAATTGC	732	Dutka-Malen et al. (1995)
	vanA$_2$	GTACAATGCGGCCGTTA		
vanB	vanB-B3-f	ACGGAATGGGAAGCCGA	647	Depardieu et al. (2004)
	vanB-B4-r	TGCACCCGATTTCGTTC		

Table 2 MIC reference values for *E. faecalis* strains for antimicrobial agents tested

Antimicrobial agent	MIC^a (µg/mL)			References
	S^b	I^b	R^b	
Amoxicillin	≤4		≥8	EUCAST (2016)[c]
Doxycycline	≤4	8	≥16	CLSI (2013)[c]
Rifampicin	≤1	2	≥4	CLSI (2013)[c]
Azithromycin	≤2		≥8	Fass (1993)
Streptomycin HL-R^d			>512	EUCAST (2016)[c]

[a]MIC = Minimum inhibitory concentration
[b]S = Susceptible, I = Intermediate, R = Resistant
[c]CLSI Clinical and Laboratory Standard Institute, EUCAST European Committee on Antimicrobial Susceptibility Testing
[d]High-level Resistance

susceptible, intermediate or resistant categories according to the EUCAST (The European Committee on Antimicrobial Susceptibility Testing) breakpoints and, where EUCAST values were not available, to the CSLI (Clinical and Laboratory Standards Institute), both listed in Table 2 (EUCAST 2016; CLSI 2013). If these standards were not available, minimal inhibitory concentration (MIC) values were compared with values for similar strains in literature.

3 DNA Isolation

Material from pure cultures was used to extract total bacterial DNA with the DNeasy Blood and

Tissue Kit (Qiagen, Hilden, Germany). The DNA extraction was performed according to the manufacturer's protocol for Gram-positive bacteria. The DNA was eluted with 200 µl AE buffer (Qiagen) and stored at −20 °C.

3.1 PCR for the Detection of *E. faecalis* Antibiotic Resistance Genes

The isolated DNA was used as a template for the amplification of nine different antibiotic resistance genes from *E. faecalis*. The primers, annealing temperatures and corresponding references are listed in Table 1. To amplify the different fragments, initial denaturation was performed at 94 °C for 5 min, followed by 35 cycles with denaturation at 94 °C for 60 s, varying annealing times (Table S1), extension at 72 °C for 60s and a final extension at 72 °C for 10 min. The primer concentration, template amount and annealing temperature varied for the different PCR systems, therefore all respective information is listed in the supplementary material (Table S1). The amplification was performed in a total volume of 25 µl and all reaction mixtures contained 1x PCR buffer (Qiagen), 0.2 mM each of the four deoxyribonucleoside triphosphates (dNTPs; PEQLAB, Erlangen, Germany) and 2.5 U Taq-Polymerase (Qiagen) and the specific amount of forward and reverse primers as well as template DNA. A no-template control and a positive control were included in each set of PCR reactions. The amplified products were visualized by gel electrophoresis using a 1% agarose gel.

4 Statistical Analysis

The correlation of the antibiotic resistance genes and the phenotypic resistance characteristics with the respective origin of the *E. faecalis* isolates was analyzed using the Fisher's exact test and pairwise comparisons were performed with the chi-square test with Bonferroni correction. The level of significance was 0.05.

5 Results

A total of 70 *E. faecalis* isolates from four different origins (endodontic, plaque/saliva, food and nosocomial isolates) were analyzed for the presence of nine antibiotic resistance genes and for their antibiotic susceptibility to five different antibiotics.

5.1 Phenotypic Antimicrobial Susceptibility of *E. faecalis* Isolates

Table S2 (Supplementary material) shows the MIC values for the tested antibiotic agents for all isolates in detail and additionally includes the resistance phenotype analyzed by VITEK in a prior study (Anderson et al. 2015). All tested isolates were susceptible to amoxicillin. As shown in Fig. 1, a high percentage of the plaque/saliva isolates from healthy individuals (81.0%), of the food (78.6%) and the nosocomial isolates (73.33%) were resistant against doxycycline, as well as against azithromycin (81.0%; 85.6% and 86.7% resp.). Rifampicin resistance was detected in about half the plaque/saliva and endodontic isolates (52.4% and 50%), in 40% of the nosocomial isolates and in a lower percentage of the food isolates (14.3%). While none of the oral isolates showed resistance against high-level streptomycin, 46.7% of the nosocomial and 35.7% of the food isolates showed resistance.

5.2 Multidrug-Resistance among *E. faecalis* Isolates from Different Origins

According to the classification recommendations suggested by the European Center for Disease Prevention and Control (ECDC) only Enterococci resistant to ≥1 agent in ≥3 antimicrobial categories relevant for these species (Magiorakos et al. 2012) can be determined as multidrug-resistant (MDR). In our study this would correspond to a combined resistance

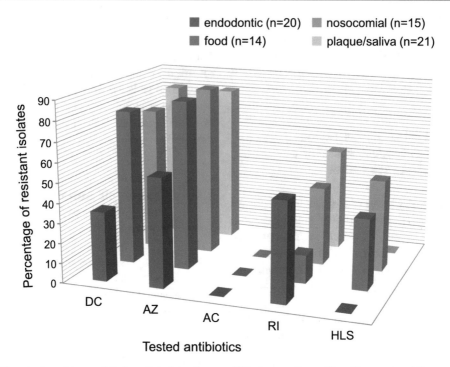

Fig. 1 Phenotypic resistance of *E. faecalis* isolates from food, secondary endodontic infections, plaque/saliva and nosocomial infections analyzed by Etest (*DC* doxycycline, *AZ* azithromycin, *AC* amoxicillin, *RI* rifampicin, *HLS* high-level streptomycin)

against high-level gentamicin, high-level streptomycin, ciprofloxacin/levofloxacin, vancomycin, linezolid and doxycycline. According to these guidelines, several isolates of the nosocomial infections (sample nr. 110053, 109891, 512359, resistant to doxycycline, high-level streptomycin, high-level gentamicin and levofloxacin/ciprofloxacin as well as 109898 and 512176, resistant to high-level streptomycin, high-level gentamicin and levofloxacin/ciprofloxacin; this study and (Anderson et al. 2015)), can be classified as multidrug-resistant. Still, many of the other tested isolates showed resistance to more than one antibiotic. Several other nosocomial isolates showed resistance to doxycycline and high-level gentamicin or high-level gentamicin and levofloxacin/ciprofloxacin. Nine plaque/saliva isolates showed combined resistance against doxycycline and high-level gentamicin and one endodontic isolate showed resistance to doxycycline and linezolid.

5.3 Distribution of Antibiotic Resistance Genes in *E. faecalis* Isolates

Figure 2 shows the percentage of the detected resistance genes in the *E. faecalis* isolates. The *tetM* gene was present in isolates from all four origins in high percentages, i.e. 65.0%, 80.0%, 86.7% and 90.5% of the endodontic, nosocomial, food and plaque/saliva isolates respectively, while other tetracycline resistance genes (*tetO*) were not detected. The *ermB* gene was detected in many nosocomial isolates (60%) and in 47.6% and 26.6% of the plaque/saliva and food isolates respectively. *Bla*$_{TEM}$ genes were found primarily in nosocomial infection isolates (53.3%), but also 30.0% of the endodontic isolates and 13.3% of the food isolates harbored these genes. *ErmC* genes and genes for the resistance to vancomycin were not present in any of the isolates.

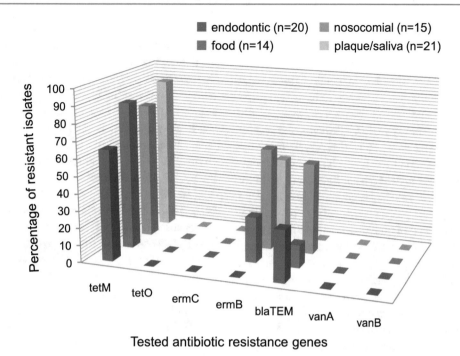

Fig. 2 Prevalence of resistance genes detected by PCR in *E. faecalis* isolates from food, secondary endodontic infections, plaque/saliva and nosocomial infections

5.4 Statistical Results – Significant Associations of Detected Resistance and the Origin of the *E. faecalis* Isolates

The analysis of possible correlations between the antibiotic resistance results and the origin of the isolates revealed that the presence of the *ermB* gene correlated with the origin (p $\leq$ 0.001), 60% of the nosocomial isolates and 47% of the plaque/saliva isolates were found positive for this gene, whereas none of the endodontic and only 26.5% of the food isolates were positive for it. The *bla*$_{TEM}$ gene also correlated with the origin (p $\leq$ 0.001), 53.3% of the nosocomial isolates possessed this trait and 30% of the endodontic isolates, yet only 13.3% of the food and none of the plaque/saliva isolates. Phenotypic doxycycline resistance correlated with the origin (p $\leq$ 0.01), it was widespread in food, nosocomial and plaque/saliva isolates (78.6%, 73.3%, 81.0%), yet only few endodontic isolates (35%)

showed resistance. Similarly, high-level streptomycin resistance correlated with the origin of the isolates (p $\leq$ 0.001), food and nosocomial isolates were frequently resistant (35.7% and 46.7% respectively), whereas all oral isolates were susceptible.

The pairwise comparison of selected traits showed significantly more doxycycline-resistant isolates from plaque/saliva and from food than from endodontic infections (p $\leq$ 0.01 and p = 0.036 resp.). Also, significantly more food isolates and nosocomial isolates were resistant to high-level streptomycin than endodontic and plaque/saliva isolates (p = 0.006 / p = 0.006 and p $\leq$ 0.001 / p $\leq$ 0.001 resp.) In addition, significantly more isolates from nosocomial infections harbored the *bla*$_{TEM}$ gene than isolates from endodontic infections (p = 0.04).

6   Discussion

Especially since the 1990s, *Enterococcus faecalis* has emerged as a leading nosocomial pathogen and has been shown to have the ability to acquire and spread resistance genes readily (Arias and Murray 2012). However, the role of the oral cavity as a potential reservoir for resistant *E. faecalis* has not been clarified yet. Therefore this study reports the antibiotic susceptibility and antibiotic resistance genes of isolates from supragingival plaque and saliva of healthy individuals and of endodontic infections in comparison to isolates from nosocomial infections and food. Notably the plaque/saliva isolates stand out regarding their equipment with tetracycline resistance genes (*tetM*) and erythromycin resistance genes (*ermB*) which are comparable to the nosocomial isolates (90.5% versus 80% and 47.6% versus 60.0% respectively). Phenotypic azithromycin resistance of the plaque/saliva isolates is similar to the nosocomial isolates and phenotypic resistance to doxycycline and rifampicin is shown in the same range in the plaque/saliva isolates as in the nosocomial isolates (81.0% versus 73.3% and 52.4% versus 40% respectively) which indicates a possible role of the oral cavity as a reservoir for these resistance traits. The doxycycline resistance in plaque/saliva was significantly higher than in the food and the endodontic isolates. Both, doxycycline and azithromycin as well as tetracycline are used for the treatment of periodontitis and other dental diseases (Poveda Roda et al. 2007; Preshaw et al. 2004; Roberts and Mullany 2010). Erythromycin, next to clindamycin, can be prescribed for patients allergic to penicillin e.g. in endodontic infections (Jacinto et al. 2003). Rifampicin is commonly used in the treatment of serious infections as well as for chemoprophylaxis in bacterial meningitis (Gaetti-Jardim et al. 2010).

Up to now, only the results of one other study analyzing plaque isolates from healthy individuals are available (Poeta et al. 2009), but the authors only detected 3 *E. faecalis* isolates among other enterococci, which were all resistant to streptomycin, erythromycin and tetracycline, possessing *ermB* and *tetM/tetL* genes. In contrast to our results, Gaetti-Jardim et al., who analyzed mixed samples of saliva and plaque from healthy individuals and those with different dental diseases found much lower resistance to doxycycline (12.9) and no resistance to rifampicin, yet 19.4% resistance to amoxicillin compared to 0% in our isolates (Gaetti-Jardim et al. 2010). This could be explained by geographical differences considering antibiotic use, since these isolates stemmed from Brazilian patients. Another very recent Brazilian study analyzed a large number of *E. faecalis* isolates from oral rinses of healthy individuals reporting a similar amoxicillin resistance (12.3%) corresponding with Gaetti-Jardims results (Komiyama et al. 2016). Fifty-three percent of the tested isolates in their study were resistant to tetracycline compared to over 90% of plaque isolates in the present study harboring the *tetM* gene, and 85.7% being phenotypically resistant in an earlier study of our group (Anderson et al. 2015).

As far as studying isolates from healthy individuals, most other authors have analyzed fecal isolates, e.g. Kuch et al. (Kuch et al. 2012), finding resistance to rifampicin (37%), to tetracycline (55.6%) and to high-level gentamicin (8.6%) mostly in the same range as our results, although slightly less than we found in plaque/saliva from healthy individuals (52.4%, 85.7% and 47.5% respectively). Lietzau *et al.* analyzed feces samples from healthy individuals in Germany and found only 29.8% of the *E. faecalis* isolates resistant to doxycycline compared to 81% of our plaque/saliva isolates (Lietzau et al. 2006). Another study that investigated a large number of nosocomial and a few commensal strains from various geographical regions detected tetracycline resistance (conferred through *tetM* and *tetL*), erythromycin resistance (conferred through *ermB*) and high level gentamicin as well as vancomycin resistance in the nosocomial isolates (McBride et al. 2007). Yet the commensal strains only harbored the *tetM* gene in contrast to our plaque/ saliva isolates of which nearly 50% possessed the ermB gene.

Endodontic isolates have been extensively analyzed for their virulence and antibiotic resistance traits, since *E. faecalis* is thought to contribute to persistent root canal infections (Anderson et al. 2013). Increasing resistances, e.g. against tetracycline, rifampicin, ciprofloxacin and erythromycin have been reported (Al-Ahmad et al. 2014). Pinheiro et al. detected 85.8% azithromycin 71.5% erythromycin and 14.3% tetracycline as well as doxycycline resistance in endodontic *E. faecalis* (Pinheiro et al. 2004). In comparison, the present study detected 55% azithromycin, 35% doxycycline and 65% tetracycline resistant isolates. A recent study by Barbosa-Ribeiro et al. detected isolates showing intermediate and full resistance to amoxicillin, azithromycin, rifampicin and doxycycline from endodontic retreatment cases, concurring with our results, although with lower percentages for the latter three antibiotics (Barbosa-Ribeiro et al. 2016). Regarding their equipment with resistance genes, our endodontic samples showed a fairly high percentage of isolates carrying the bla_{TEM} gene (30%) compared with our nosocomial isolates (53.3%). Although enterococci possess intrinsic resistance against beta-lactam antibiotics, this resistance varies and e.g. ampicillin still has a high effectiveness against *E. faecalis* (Kristich et al. 2014). Other authors (Rocas and Siqueira 2013; Jungermann et al. 2011) have found bla_{TEM} in DNA extracts from endodontic infections. The equipment with resistance genes as well as phenotypic resistance against other tested antibiotics was lower for the endodontic isolates in the present study than for isolates of other sources.

For the last two decades most studies of nosocomial *E. faecalis* isolates regarding their antibiotic resistance have focused specifically on vancomycin resistance, while data on other antibiotics is less frequent (Ruiz-Garbajosa et al. 2006). In our study, we did not detect any vancomycin resistance in the nosocomial isolates, nor the isolates from other origins, which for isolates from the oral cavity is a consistent and favorable finding, considering the function of vancomycin as reserve antibiotic (Barbosa-Ribeiro et al. 2016; Komiyama et al. 2016; Pinheiro et al. 2004; Rams et al. 2013; C. M. Sedgley et al. 2005a; Dahlen et al. 2012).

On the other hand, high percentages of the nosocomial isolates of the present study exhibited doxycycline and azithromycin resistance as well as the possession of bla_{TEM}, *tetM* and *ermB*. Azithromycin resistance has also been found in a high percentage of nosocomial isolates from German patients by Wenzler *et al.* (72% versus 86.6% percent in our study) (Wenzler et al. 2004) and frequent rifampicin resistance has been reported by Kuch *et al.* (36.7% versus 40% in our study) (Kuch et al. 2012). Kuch also reported high-level streptomycin resistance in 26.7% of the German nosocomial isolates (Kuch et al. 2012) which is somewhat higher in our study, with 46.7% of our nosocomial isolates showing resistance.

Regarding the food isolates tested in the present study, we found a high resistance to doxycycline and azithromycin, comparable to the plaque/ saliva and nosocomial isolates, as well as frequent possession of *tetM*, while they were susceptible to most other tested antibiotics. In contrast to the oral isolates those from food showed a measure of high-level streptomycin resistance (35.7%). *E. faecalis* is frequently detected not only in raw milk but also raw milk cheeses (Jamet et al. 2012) and isolates have been found to be resistant to various antibiotics, e.g. tetracycline, erythromycin, rifampicin and streptomycin in concordance with our results (Jamet et al. 2012; Schlegelova et al. 2002; Templer and Baumgartner 2007). Jamet et al. found *tetM* and *ermB* genes widespread in multiple resistant cheese isolates, and their multilocus sequence typing (MLST) analysis revealed several isolates from clonal complexes that have been associated with periodontitis (Jamet et al. 2012). This finding is significant against the backdrop of the study of Al-Ahmad et al. reporting that *E. faecalis* isolates from cheese were able to integrate into the oral biofilm *in vivo* (Al-Ahmad et al. 2010). Thus, also food isolates, e.g. raw milk and products thereof could serve as a resistance reservoir and facilitate the spread of resistance through transfer of genes to oral *E. faecalis*. The possibility of a transfer of resistance genes from these cheese isolates to the oral isolates or among different oral *E. faecalis* isolates through conjugative transposons *in vivo* has not been studied yet. Nevertheless, oral *E. faecalis* has been proven to be a recipient of resistance genes from other species (Roberts et al. 2001) *in vitro* and transfer of plasmid coded erythromycin resistance has been shown in root canals *ex vivo* (C. M. Sedgley et al. 2008). The presence of the Tn916-like element, carrying e.g. *tetM*, and other transposons has been detected in *E. faecalis* from food as well as oral enterococci (Jamet et al. 2012; Roberts and Mullany 2011; Kristich et al. 2014).

In summary we can conclude that both the oral cavity, in particular dental plaque and saliva, as well as foods can present a reservoir of *E. faecalis* strains with multiple antibiotic resistances including the potential of resistance transfer to other strains or even other species. Consequently, continued monitoring of *E. faecalis* for antibiotic resistance should be performed not only for nosocomial, but also for oral strains.

Funding Information This study was partly supported by the German Research foundation (DFG, AL-1179/2–1) and by the Ministry of Science, Research and the Arts in Baden-Württemberg (VR-MED_FAK).

Acknowledgements The authors thank Bettina Spitzmüller, Kristina Kollmar and Annette Wittmer for excellent technical assistance and Nicole Arweiler, Daniel Jonas and Ingrid Huber for providing part of the isolates, as well as Grant Anderson for English language correction.

Conflicts of Interest The authors deny any conflicts of interest related to this study.

Ethical Statement All endodontic and clinical isolates were obtained after approval by the Ethics Committee (no. 140/09, University of Freiburg).

Supplementary Material

Table S1 PCR amplification and cycling conditions for the detected antibiotic resistance genes

Target	Primer Concentrations	MgCl$_2$[a]	Template DNA	Annealing
E. faecalis				
bla$_{TEM}$	1.0 µl (5 µM) each	1 µl (2 mM)	1 µl	60 °C 30 s
tet(M)	0.5 µl (5 µM) each	–	1 µl	58 °C 45 s
tet(O)	0.5 µl (5 µM) each	–	2 µl	53 °C 30 s
erm(B)	0.5 µl (5 µM) each	–	2 µl	52 °C 30 s
erm(C)	0.25 µl (5 µM) each	–	2 µl	54 °C 60 s
*van*A	1.0 µl (5 µM) each	–	1 µl	54 °C 60 s
*van*B	2.0 µl (5 µM) each	–	1 µl	54 °C 60 s

[a]Qiagen, Hilden, Germany

Table S2 Antibiotic susceptibility and presence of antibiotic resistance genes in 70 *Enterococcus faecalis* isolates from four different sources

Isolates	Antibiotic resistance genes[a]	MIC (µg/mL)[b]					Resistance phenotype	Resistance phenotype[d] VITEK (Anderson 2015)
		DC[c]	AZ[c]	AC[c]	RI[c]	HLS[c]		
Endodontic								
1aR1		0.5	3	0.75	1	64		ERY-*T*/*S*
1anR8	*bla*TEM	**16**	**12**	0.75	**6**	128	DC-AZ-RI	ERY-*T*/*S*
11aRSP		0.75	2	0.75	3	128		ERY-*T*/*S*
12aSP	*tetM*	**24**	**8**	0.5	**> 32**	96	DC-AZ-RI	ERY-*T*/*S*-TET
17aSP	*tetM*	**16**	**24**	0.38	**4**	32	DC-AZ-RI	ERY- *T*/*S*-TET-LIN
21aSP	*tetM*	**24**	**12**	0.75	**8**	128	DC-AZ	ERY-*T*/*S*-TET
33aR8	*bla*TEM	**16**	**8**	0.75	2	64	DC-AZ	ERY-*T*/*S*
44aR6	*bla*$_{TEM-1}$, *tetM*	6	**8**	0.5	3	64	AZ	ERY-*T*/*S*-TET
44aREnA	*tetM*	3	6	0.75	3	64		ERY-*T*/*S*-TET
44aF6	*bla*$_{TEM-1}$, *tetM*	3	**8**	0.75	3	64	AZ	ERY-*T*/*S*-TET
44aFEnA	*bla*$_{TEM-1}$, *tetM*	3	6	0.75	3	64		ERY-*T*/*S*-TET
44anR7	*tetM*	3	6	0.5	**4**	64	RI	ERY-*T*/*S*-TET
44anR10	*tetM*	3	6	0.75	**4**	64	RI	ERY-*T*/*S*-TET
44anF7	*tetM*	4	**8**	0.5	**4**	64	AZ-RI	ERY-*T*/*S*-TET
45aSP7	*bla*$_{TEM-1}$	0.75	3	0.75	**6**	64	RI	ERY-*T*/*S*
RGFR-81G8		0.38	**48**	0.75	**32**	192	AZ-RI	n.d.
RG20R72C3	*tetM*	8	**> 256**	0.38	1	96	AZ	n.d.
RG18F102F2	*tetM*	**16**	1.5	0.5	**> 32**	128	DC-RI	ERY-*T*/*S*-TET
MFCT7501C6		1	6	0.5	0.5	48		ERY-*T*/*S*
MFCT23S01A1	*tetM*	**24**	**12**	0.5	**4**	96	DC-AZ-RI	ERY-*T*/*S*-TET
Plaque/Saliva								
90sp		1	**8**	1	**> 32**	96	AZ-RI	ERY-*T*/*S*
223sp	*tetM*	1.5	1	0.5	0.38	128		ERY-*T*/*S*
254p	*tetM*, *ermB*	**32**	**> 256**	1.5	3	96	DC-AZ	ERY-*T*/*S*-TET-HGEN
255p	*tetM*	**24**	6	0.5	**8**	96	DC-RI	ERY-*T*/*S*-TET
282sp	*tetM*, *ermB*	**24**	**> 256**	1	3	96	DC-AZ	ERY-*T*/*S*-TET-HGEN
288p	*tetM*, *ermB*	**16**	**> 256**	2	3	96	DC-AZ	ERY-*T*/*S*-TET-HGEN
291sp	*tetM*, *ermB*	**16**	**> 256**	1.5	2	96	DC-AZ	ERY-*T*/*S*-TET-HGEN
294sp	*tetM*, *ermB*	**16**	**> 256**	1	**4**	96	DC-AZ	ERY-*T*/*S*-TET-HGEN
319p	*tetM*, *ermB*	**24**	**> 256**	1	3	96	DC-AZ	ERY-*T*/*S*-TET-HGEN
327p	*tetM*, *ermB*	**16**	**> 256**	1.5	2	96	DC-AZ	ERY-*T*/*S*-TET-HGEN
351p	*tetM*	**16**	6	1	**6**	96	DC-AZ-RI	ERY-*T*/*S*-TET
353p	*tetM*	**24**	6	0.5	**6**	96	DC-AZ-RI	ERY-*T*/*S*-TET
354p	*tetM*	**16**	6	0.75	**16**	96	DC-AZ-RI	ERY-*T*/*S*-TET
357sp		0.5	1.5	0.75	2	128		ERY-*T*/*S*

(continued)

Table S2 (continued)

Isolates	Antibiotic resistance genes[a]	MIC (µg/mL)[b]					Resistance phenotype	Resistance phenotype[d] VITEK (Anderson 2015)
		DC[c]	AZ[c]	AC[c]	RI[c]	HLS[c]		
359sp	*tetM*	**16**	**8**	0.75	**> 32**	96	DC-AZ-RI	ERY-*T*/*S*-TET
383sp	*tetM*	8	**8**	0.75	**> 32**	128	AZ-RI	ERY-*T*/*S*-TET
446sp	*tetM, ermB*	**24**	**> 256**	1.5	**4**	96	DC-AZ-RI	ERY-*T*/*S*-TET-HGEN
447sp	*tetM, ermB*	**16**	**> 256**	0.75	**4**	96	DC-AZ-RI	ERY-*T*/*S*-TET-HGEN
452sp	*tetM, ermB*	**16**	**> 256**	0.75	3	96	DC-AZ	ERY-*T*/*S*-TET-HGEN
478sp	*tetM*	**24**	**8**	0.75	**4**	48	DC-AZ-RI	ERY-*T*/*S*-TET
513sp	*tetM*	**16**	6	0.75	**12**	64	DC-RI	ERY-*T*/*S*-TET
Food								
F2/19	*bla$_{TEM-1}$, tetM*	12	6	1	1.5	**> 1024**	HLS	ERY-*T*/*S*-TET
E392	*tetM*	**16**	**12**	1.5	0.75	96	DC-AZ	ERY-*T*/*S*-TET
C339	*tetM, ermB*	6	**96**	0.75	1.5	96	AZ	ERY-*T*/*S*-TET
C350	*tetM, ermB*	**32**	**> 256**	0.75	1.5	**> 1024**	DC-AZ-HLS	ERY-*T*/*S*-TET
C409	*tetM*	**64**	**16**	0.75	1.5	**> 1024**	DC-AZ-HLS	ERY-*T*/*S*-TET
C528		0.5	3	0.75	1	128		ERY-*T*/*S*
C671	*tetM*	**16**	**16**	0.75	3	192	DC-AZ	ERY-*T*/*S*-TET
C686	*tetM*	**16**	**16**	0.75	**4**	96	DC-AZ-RI	ERY-*T*/*S*-TET
C725/3	*tetM*	**16**	**24**	0.75	2	96	DC-AZ	ERY-*T*/*S*-TET
C729	*tetM, ermB*	**24**	**> 256**	1	0.75	**> 1024**	DC-AZ-HLS	ERY-*T*/*S*-TET
C737/1	*tetM, ermB*	**64**	**> 256**	0.75	0.75	64	DC-AZ	ERY-*T*/*S*-TET
C771	*tetM*	**16**	**8**	0.75	1	96	DC-AZ	ERY-*T*/*S*-TET
C890	*bla$_{TEM-1}$, tetM*	**16**	**16**	0.75	1	96	DC-AZ	ERY-*T*/*S*-TET
C906/1	*tetM*	**16**	**16**	1	**4**	**> 1024**	DC-AZ-RI-HLS	ERY-*T*/*S*-TET
Nosocomial								
110028	*bla$_{TEM-1}$, tetM*	**24**	**8**	0.75	2	96	DC-AZ	ERY-*T*/*S*-TET
110035	*tetM, erm*B	**24**	**> 256**	0.75	**8**	128	DC-AZ-RI	ERY-T/S-TET-HGEN
110047		0.38	4	0.75	**8**	96	DC-AZ-RI	ERY-*T*/*S*
110053[e]	*bla$_{TEM-1}$, tetM, ermB*	**16**	**> 256**	1	1.5	**> 1024**	DC-AZ-HLS	ERY-T/S-TET-HGEN-LEV-CIP
109891	*bla$_{TEM-1}$, tetM, ermB*	**24**	**> 256**	1.5	0.75	**> 1024**	DC-AZ-HLS	ERY-T/S-TET-HGEN-LEV-CIP
109898	*bla$_{TEM-1}$, tetM, ermB*	**24**	**> 256**	0.75	**4**	**> 1024**	DC-AZ-RI-HLS	ERY-*T*/*S*-TET-HGEN
229355	*bla$_{TEM-1}$*	4	**8**	0.5	2	128	AZ	ERY-*T*/*S*-TET
512106	*tetM*	**24**	**16**	0.5	**4**	96	DC-AZ-RI	ERY-*T*/*S*-TET
512118	*bla$_{TEM-1}$, tetM, ermB*	**24**	**> 256**	1	0.75	**> 1024**	DC-AZ-HLS	ERY-T/S-TET

(continued)

Table S2 (continued)

Isolates	Antibiotic resistance genes[a]	MIC (µg/mL)[b]					Resistance phenotype	Resistance phenotype[d] VITEK (Anderson 2015)
		DC[c]	AZ[c]	AC[c]	RI[c]	HLS[c]		
512129	*tetM*	**16**	3	1	> 32	> 1024	DC-RI-HLS	ERY-*T*/*S*-TET
512176	*bla*$_{TEM-1}$, *ermB*	0.75	> 256	0.75	0.75	> 1024	AZ-HLS	ERY-T/S-HGEN-LEV-CIP
512188	*tetM*, *ermB*	**24**	> 256	0.5	> 32	96	AZ-RI	ERY-T/S-TET-HGEN-LEV-CIP
512276	*tetM*, *ermB*	**16**	> 256	1	1.5	96	DC-AZ	ERY-*T*/*S*-TET-HGEN
512298	*tetM*	6	**8**	0.75	2	96	DC-AZ	ERY-*T*/*S*-TET
512359	*bla*$_{TEM-1}$,*tetM*, *ermB*	**64**	> 256	1	1.5	> 1024	DC-AZ-HLS	ERY-T/S-TET-HGEN-LEV-CIP

[a]Genes detected: *bla*$_{TEM-1}$, *tetM*, *tetW*, *tetQ*, *ermC*, *ermB*, *vanA*, *vanB*, *bla*$_{TEM}$
[b]minimal inhibitory concentrations, resistant isolates are marked in bold
[c]doxycycline (DC); azithromycin (AZ); amoxicillin (AC); rifampicin (RI); high-level-streptomycin (HLS)
[d]Resistance phenotypes analyzed with the VITEK-System, results published in previous study (Anderson et al. 2015); Erythromycin (ERY); tetracycline (TET); high-level gentamicin (HGEN); trimethoprim/sulfamethoxazol (T/S), *(T/S)* = *intermediate*; linezolid (LIN); levofloxacin (LEV); ciprofloxacin (CIP)
[e]bold: multi-drug-resistant isolates

References

Al-Ahmad A, Maier J, Follo M, Spitzmuller B, Wittmer A, Hellwig E et al (2010) Food-borne enterococci integrate into oral biofilm: an in vivo study. J Endod 36(11):1812–1819. doi:10.1016/j.joen.2010.08.011

Al-Ahmad A, Ameen H, Pelz K, Karygianni L, Wittmer A, Anderson AC et al (2014) Antibiotic resistance and capacity for biofilm formation of different bacteria isolated from endodontic infections associated with root-filled teeth. J Endod 40(2):223–230. doi:10.1016/j.joen.2013.07.023

Amsler K, Santoro C, Foleno B, Bush K, Flamm R (2010) Comparison of broth microdilution, agar dilution, and Etest for susceptibility testing of doripenem against gram-negative and gram-positive pathogens. J Clin Microbiol 48(9):3353–3357. doi:10.1128/jcm.00494-10

Anderson AC, Al-Ahmad A, Elamin F, Jonas D, Mirghani Y, Schilhabel M et al (2013) Comparison of the bacterial composition and structure in symptomatic and asymptomatic endodontic infections associated with root-filled teeth using pyrosequencing. PLoS One 8(12):e84960. doi:10.1371/journal.pone.0084960

Anderson AC, Jonas D, Huber I, Karygianni L, Wolber J, Hellwig E et al (2015) Enterococcus faecalis from food, clinical specimens, and oral sites: prevalence of virulence factors in association with biofilm formation. Front Microbiol 6:1534. doi:10.3389/fmicb.2015.01534

Arias CA, Murray BE (2012) The rise of the Enterococcus: beyond vancomycin resistance. Nat Rev Microbiol 10(4):266–278. doi:10.1038/nrmicro2761

Barbosa-Ribeiro M, De-Jesus-Soares A, Zaia AA, Ferraz CC, Almeida JF, Gomes BP (2016) Antimicrobial susceptibility and characterization of virulence genes of Enterococcus faecalis isolates from teeth with failure of the endodontic treatment. J Endod. doi:10.1016/j.joen.2016.03.015

Call DR, Bakko MK, Krug MJ, Roberts MC (2003) Identifying antimicrobial resistance genes with DNA microarrays. Antimicrob Agents Chemother 47(10):3290–3295

CLSI (2013) CLSI. Performance Standards for Antimicrobial Susceptibility Testing; Twenty-Third Informational Supplement. CLSI document M100-S23. Clinical and Laboratory Standards Institute, Wayne, PA

Dahlen G, Samuelsson W, Molander A, Reit C (2000) Identification and antimicrobial susceptibility of enterococci isolated from the root canal. Oral Microbiol Immunol 15(5):309–312

Dahlen G, Blomqvist S, Almstahl A, Carlen A (2012) Virulence factors and antibiotic susceptibility in enterococci isolated from oral mucosal and deep infections. J Oral Microbiol:4. doi:10.3402/jom.v4i0.10855

Depardieu F, Kolbert M, Pruul H, Bell J, Courvalin P (2004) VanD-type vancomycin-resistant Enterococcus faecium and Enterococcus faecalis. Antimicrob Agents Chemother 48(10):3892–3904. doi:10.1128/aac.48.10.3892-3904.2004

Duggan JM, Sedgley CM (2007) Biofilm formation of oral and endodontic Enterococcus faecalis. J Endod 33(7):815–818. doi:10.1016/j.joen.2007.02.016

Dutka-Malen S, Evers S, Courvalin P (1995) Detection of glycopeptide resistance genotypes and identification to the species level of clinically relevant enterococci by PCR. J Clin Microbiol 33(1):24–27

EUCAST (2016) The European Committee on Antimicrobial Susceptibility Testing. Breakpoint tables for interpretation of MICs and zone diameters. Version 6.0, 2016 http://www.eucast.org. http://www.eucast.org/clinical_breakpoints/

Fass RJ (1993) Erythromycin, clarithromycin, and azithromycin: use of frequency distribution curves, scattergrams, and regression analyses to compare in vitro activities and describe cross-resistance. Antimicrob Agents Chemother 37(10):2080–2086

Fisher K, Phillips C (2009) The ecology, epidemiology and virulence of Enterococcus. Microbiology 155 (Pt 6):1749–1757. doi:10.1099/mic.0.026385-0

Franz CM, Huch M, Abriouel H, Holzapfel W, Galvez A (2011) Enterococci as probiotics and their implications in food safety. Int J Food Microbiol 151 (2):125–140. doi:10.1016/j.ijfoodmicro.2011.08.014

Gaetti-Jardim EC, Marqueti AC, Faverani LP, Gaetti-Jardim E Jr (2010) Antimicrobial resistance of aerobes and facultative anaerobes isolated from the oral cavity. J Appl Oral Sci 18(6):551–559

Hammerum AM (2012) Enterococci of animal origin and their significance for public health. Clin Microbiol Infect 18(7):619–625. doi:10.1111/j.1469-0691.2012.03829.x

Jacinto RC, Gomes BP, Ferraz CC, Zaia AA, Filho FJ (2003) Microbiological analysis of infected root canals from symptomatic and asymptomatic teeth with periapical periodontitis and the antimicrobial susceptibility of some isolated anaerobic bacteria. Oral Microbiol Immunol 18(5):285–292

Jamet E, Akary E, Poisson MA, Chamba JF, Bertrand X, Serror P (2012) Prevalence and characterization of antibiotic resistant Enterococcus faecalis in French cheeses. Food Microbiol 31(2):191–198. doi:10.1016/j.fm.2012.03.009

Jorgensen JH, Ferraro MJ (2009) Antimicrobial susceptibility testing: a review of general principles and contemporary practices. Clin Infect Dis 49 (11):1749–1755. doi:10.1086/647952

Jungermann GB, Burns K, Nandakumar R, Tolba M, Venezia RA, Fouad AF (2011) Antibiotic resistance in primary and persistent endodontic infections. J Endod 37(10):1337–1344. doi:10.1016/j.joen.2011.06.028

Komiyama EY, Lepesqueur LS, Yassuda CG, Samaranayake LP, Parahitiyawa NB, Balducci I et al (2016) Enterococcus species in the oral cavity: prevalence, virulence factors and antimicrobial susceptibility. PLoS One 11(9):e0163001. doi:10.1371/journal.pone.0163001

Kouidhi B, Zmantar T, Mahdouani K, Hentati H, Bakhrouf A (2011) Antibiotic resistance and adhesion properties of oral Enterococci associated to dental caries. BMC Microbiol 11:155. doi:10.1186/1471-2180-11-155

Kristich CJ, Rice LB, Arias CA (2014) Enterococcal infection-treatment and antibiotic resistance. In: Gilmore MS, Clewell DB, Ike Y, Shankar N (eds) Enterococci: from commensals to leading causes of drug resistant infection. Massachusetts Eye and Ear Infirmary, Boston

Kuch A, Willems RJ, Werner G, Coque TM, Hammerum AM, Sundsfjord A et al (2012) Insight into antimicrobial susceptibility and population structure of contemporary human Enterococcus faecalis isolates from Europe. J Antimicrob Chemother 67(3):551–558. doi:10.1093/jac/dkr544

Lietzau S, Hoewner M, von Baum H, Marre R, Brenner H (2006) Antibiotic resistant fecal isolates of Enterococci among unselected patients outside the clinical sector: an epidemiological study from Southern Germany. Pharmacoepidemiol Drug Saf 15 (4):275–277. doi:10.1002/pds.1167

Magiorakos AP, Srinivasan A, Carey RB, Carmeli Y, Falagas ME, Giske CG et al (2012) Multidrug-resistant, extensively drug-resistant and pandrug-resistant bacteria: an international expert proposal for interim standard definitions for acquired resistance. Clin Microbiol Infect 18(3):268–281. doi:10.1111/j.1469-0691.2011.03570.x

Manson JM, Hancock LE, Gilmore MS (2010) Mechanism of chromosomal transfer of Enterococcus faecalis pathogenicity island, capsule, antimicrobial resistance, and other traits. Proc Natl Acad Sci U S A 107(27):12269–12274. doi:10.1073/pnas.1000139107

McBride SM, Fischetti VA, Leblanc DJ, Moellering RC Jr, Gilmore MS (2007) Genetic diversity among Enterococcus faecalis. PLoS One 2(7):e582. doi:10.1371/journal.pone.0000582

Murray BE (1990) The life and times of the Enterococcus. Clin Microbiol Rev 3(1):46–65

Paganelli FL, Willems RJ, Leavis HL (2012) Optimizing future treatment of enterococcal infections: attacking the biofilm? Trends Microbiol 20(1):40–49. doi:10.1016/j.tim.2011.11.001

Palmer KL, Kos VN, Gilmore MS (2010) Horizontal gene transfer and the genomics of enterococcal antibiotic resistance. Curr Opin Microbiol 13(5):632–639. doi:10.1016/j.mib.2010.08.004

Paulsen IT, Banerjei L, Myers GS, Nelson KE, Seshadri R, Read TD et al (2003) Role of mobile DNA in the evolution of vancomycin-resistant Enterococcus faecalis. Science 299(5615):2071–2074. doi:10.1126/science.1080613

Perez-Trallero E, Montes M, Orden B, Tamayo E, Garcia-Arenzana JM, Marimon JM (2007) Phenotypic and genotypic characterization of Streptococcus pyogenes isolates displaying the MLSB phenotype of macrolide resistance in Spain, 1999 to 2005. Antimicrob Agents Chemother 51(4):1228–1233. doi:10.1128/aac.01054-06

Perreten V, Vorlet-Fawer L, Slickers P, Ehricht R, Kuhnert P, Frey J (2005) Microarray-based detection of 90 antibiotic resistance genes of gram-positive bacteria. J Clin Microbiol 43(5):2291–2302. doi:10.1128/jcm.43.5.2291-2302.2005

Pinheiro ET, Gomes BP, Drucker DB, Zaia AA, Ferraz CC, Souza-Filho FJ (2004) Antimicrobial susceptibility of Enterococcus faecalis isolated from canals of root filled teeth with periapical lesions. Int Endod J 37 (11):756–763. doi:10.1111/j.1365-2591.2004.00865.x

Poeta P, Igrejas G, Goncalves A, Martins E, Araujo C, Carvalho C et al (2009) Influence of oral hygiene in patients with fixed appliances in the oral carriage of antimicrobial-resistant *Escherichia coli* and Enterococcus isolates. Oral Surg Oral Med Oral Pathol Oral Radiol Endod 108(4):557–564. doi:10.1016/j.tripleo.2009.06.002

Portenier I, Waltimo TMT, Haapasalo M (2003) Enterococcus faecalis– the root canal survivor and 'star' in post-treatment disease. Endod Top 6(1):135–159. doi:10.1111/j.1601-1546.2003.00040.x

Poveda Roda R, Bagan JV, Sanchis Bielsa JM, Carbonell Pastor E (2007) Antibiotic use in dental practice: a review. Med Oral Patol Oral Cir Bucal 12(3):E186–E192

Preshaw PM, Hefti AF, Jepsen S, Etienne D, Walker C, Bradshaw MH (2004) Subantimicrobial dose doxycycline as adjunctive treatment for periodontitis: a review. J Clin Periodontol 31(9):697–707. doi:10.1111/j.1600-051X.2004.00558.x

Rams TE, Feik D, Mortensen JE, Degener JE, van Winkelhoff AJ (2013) Antibiotic susceptibility of periodontal Enterococcus faecalis. J Periodontol 84 (7):1026–1033. doi:10.1902/jop.2012.120050

Reinert RR, Filimonova OY, Al-Lahham A, Grudinina SA, Ilina EN, Weigel LM et al (2008) Mechanisms of macrolide resistance among Streptococcus pneumoniae isolates from Russia. Antimicrob Agents Chemother 52(6):2260–2262. doi:10.1128/aac.01270-07

Roberts AP, Mullany P (2010) Oral biofilms: a reservoir of transferable, bacterial, antimicrobial resistance. Expert Rev Anti-Infect Ther 8(12):1441–1450. doi:10.1586/eri.10.106

Roberts AP, Mullany P (2011) Tn916-like genetic elements: a diverse group of modular mobile elements conferring antibiotic resistance. FEMS Microbiol Rev 35(5):856–871. doi:10.1111/j.1574-6976.2011.00283.x

Roberts AP, Cheah G, Ready D, Pratten J, Wilson M, Mullany P (2001) Transfer of TN916-like elements in microcosm dental plaques. Antimicrob Agents Chemother 45(10):2943–2946. doi:10.1128/aac.45.10.2943-2946.2001

Rocas IN, Siqueira JF Jr (2013) Detection of antibiotic resistance genes in samples from acute and chronic endodontic infections and after treatment. Arch Oral Biol 58(9):1123–1128. doi:10.1016/j.archoralbio.2013.03.010

Ruiz-Garbajosa P, Canton R, Pintado V, Coque TM, Willems R, Baquero F et al (2006) Genetic and phenotypic differences among Enterococcus faecalis clones from intestinal colonisation and invasive disease. Clin Microbiol Infect 12(12):1193–1198. doi:10.1111/j.1469-0691.2006.01533.x

Schirrmeister JF, Liebenow AL, Braun G, Wittmer A, Hellwig E, Al-Ahmad A (2007) Detection and eradication of microorganisms in root-filled teeth associated with periradicular lesions: an in vivo study. J Endod 33(5):536–540. doi:10.1016/j.joen.2007.01.012

Schlegelova J, Babak V, Klimova E, Lukasova J, Navratilova P, Sustackova A et al (2002) Prevalence of and resistance to anti-microbial drugs in selected microbial species isolated from bulk milk samples. J Vet Med B Infect Dis Vet Public Health 49 (5):216–225

Sedgley CM, Molander A, Flannagan SE, Nagel AC, Appelbe OK, Clewell DB et al (2005a) Virulence, phenotype and genotype characteristics of endodontic Enterococcus spp. Oral Microbiol Immunol 20 (1):10–19. doi:10.1111/j.1399-302X.2004.00180.x

Sedgley CM, Nagel AC, Shelburne CE, Clewell DB, Appelbe O, Molander A (2005b) Quantitative real-time PCR detection of oral Enterococcus faecalis in humans. Arch Oral Biol 50(6):575–583. doi:10.1016/j.archoralbio.2004.10.017

Sedgley C, Buck G, Appelbe O (2006) Prevalence of Enterococcus faecalis at multiple oral sites in endodontic patients using culture and PCR. J Endod 32 (2):104–109. doi:10.1016/j.joen.2005.10.022

Sedgley CM, Lee EH, Martin MJ, Flannagan SE (2008) Antibiotic resistance gene transfer between Streptococcus gordonii and Enterococcus faecalis in root canals of teeth ex vivo. J Endod 34(5):570–574. doi:10.1016/j.joen.2008.02.014

Siqueira JF Jr, Rocas IN (2004) Polymerase chain reaction-based analysis of microorganisms associated with failed endodontic treatment. Oral Surg Oral Med Oral Pathol Oral Radiol Endod 97(1):85–94. doi:10.1016/s1079210403003536

Souto R, Colombo AP (2008) Prevalence of Enterococcus faecalis in subgingival biofilm and saliva of subjects with chronic periodontal infection. Arch Oral Biol 53 (2):155–160. doi:10.1016/j.archoralbio.2007.08.004

Sun J, Song X, Kristiansen BE, Kjaereng A, Willems RJ, Eriksen HM et al (2009) Occurrence, population structure, and antimicrobial resistance of enterococci in marginal and apical periodontitis. J Clin Microbiol 47(7):2218–2225. doi:10.1128/jcm.00388-09

Templer SP, Baumgartner A (2007) Enterococci from Appenzeller and Schabziger raw milk cheese: antibiotic resistance, virulence factors, and persistence of particular strains in the products. J Food Prot 70 (2):450–455

Tenover FC, Baker CN, Swenson JM (1996) Evaluation of commercial methods for determining antimicrobial susceptibility of Streptococcus Pneumoniae. J Clin Microbiol 34(1):10–14

Thurnheer T, Belibasakis GN (2015) Integration of non-oral bacteria into in vitro oral biofilms. Virulence 6(3):258–264. doi:10.4161/21505594.2014.967608

Van Tyne D, Gilmore MS (2014) Friend turned foe: evolution of enterococcal virulence and antibiotic resistance. Annu Rev Microbiol 68:337–356. doi:10.1146/annurev-micro-091213-113003

Wenzler S, Schmidt-Eisenlohr E, Daschner F (2004) Comparative in vitro activities of three new quinolones and azithromycin against aerobic pathogens causing respiratory tract and abdominal wound infections. Chemotherapy 50(1):40–42. doi:10.1159/000077284

Adv Exp Med Biol - Advances in Microbiology, Infectious Diseases and Public Health (2018) 9: 63–72
DOI 10.1007/5584_2017_54
© Springer International Publishing AG 2017
Published online: 22 June 2017

Veterinary Public Health in Italy: From Healthy Animals to Healthy Food, Contribution to Improve Economy in Developing Countries

Margherita Cacaci and Rossella Colomba Lelli

Abstract

The role of the veterinarian as a public health officer is intrinsic to the history and the culture of veterinary organization in Italy. The Veterinary service being part of the Health administration since the birth of the Italian State in the XIX Century. In the second half of the last century the birth of the Italian National Health Service confirmed that the function of the Italian veterinary service was to analyze and reduce the risks for the human population connected to the relationship man-animal-environment, animal health, food safety and security. The Italian Veterinary Medicine School *curricula,* reflected this "model" of veterinarian as well. In the majority of countries in the world, Veterinary Services are organized within the Agriculture Administration with the main function to assure animal health and wellbeing. After the so-called "Mad-cow crisis" the awareness of the direct and essential role of veterinary services in the prevention of human illness has been officially recognized and in the third millennium the old concept of "one health" and "human-animal interface" has gained popularity worldwide.

The concept of Veterinary Public Health, has evolved at International level and has incorporated the more than a century old vision of the Italian Veterinary medicine and it is defined as "the sum of the contributions to the physical, mental and social development of people through the knowledge and application of veterinary science" (WHO, Future trends in veterinary public health. Gruppo di lavoro OMS: TE, Italy, 1999, Available from: http://www.who.int/zoonoses/vph/en/. Last visited 16 Feb 2016, 1999).

On the subject of Cooperation, Sustainability and Public Health, the EXPO 2015 event and the activities of international organizations WHO, FAO and World Organization for Animal Health are refocusing at present their worldwide mandate to protect human health and the economy of both the poorest Countries and the developed countries, according to the "new" concept of Veterinary Public Health.

M. Cacaci
Institute of Microbiology, Catholic University of the Sacred Heart (UCSC), Rome, Italy
e-mail: margherita.c86@gmail.com

R.C. Lelli (✉)
Istituto Zooprofilattico Sperimentale dell'Abruzzo e del Molise "G. Caporale", Teramo, Italy
e-mail: rossella.silmar@gmail.com

Focus of Italian Veterinary Services activity is connected to research, diagnosis and epidemiological analysis of infectious diseases, including zoonosis, food safety as well as food security.

Keywords

Cooperation · Food safety · Food security · Public Health · Veterinary services · Zoonosis

1   Introduction

Following the studies carried out by the World Health Organization (WHO), which indicate that in 2050 the population growth on Earth will exceed 9 billion people, the Food and Agriculture Organization (FAO) and the United Nations have estimated that to meet the needs regarding availability of food, the agricultural production must be more than doubled, taking also into account economic development, with special reference to some Member States (UN 2015).

The necessary increase in food production, will have a sustainable impact on environment and on the availability of natural resources. A new approach regarding the multidisciplinary collaboration between the different actors and the fallout of the results on world population is required. Furthermore, as the population density is growing, it is important to take into account the environmental impact of migration, urbanization, pollution, limited availability of natural resources, climate change and their consequences, including health problems, such as the events related to "emerging/re-emerging" diseases, including zoonoses, food borne diseases and food safety and security. In recent years, with regard to those arguments, reference is made to the slogan "One World, One Health, One Medicine", which today is evolving to "One Medicine, One Science" (Travis et al. 2014), and, more important, the concept of "One Health"(Sikkema and Koopmans 2016) is accompanied by the terms "Eco health" and "Ecosystem health" (Zinsstag 2013; Zinsstag et al. 2015; One Health Eco Health Conference 2016).

In Italy, Veterinary Public Health (VPH), related to International Cooperation and contribution to the countries' economy, is an aspect that Veterinary Medicine has taken into consideration since its beginning; the *curricula* of Veterinary Schools is based on that assumption. The concept of VPH has evolved over time, and today WHO defines it as "the sum of the contributions to physical, mental and social development of people through the knowledge and application of veterinary science"(WHO 1999).

In Italy, in 1997, Marabelli and Mantovani have defined VPH as the sum of "actions that the consumers and public administration expect from Veterinary Medicine (especially from the Veterinary State Services) for health protection, economy, environment and interrelation with animals" (Baldelli 2012).

This means that the Italian concept of VPH embraces all veterinary activities of public relevance and all the activities of Public Veterinary Services. To confirm this, it is worth considering that Veterinary Services are part of Public Health Administration, dedicated to the protection of human and animal health. In European Union, Austria is the only Member State, besides Italy, where Veterinary Services are historically part of Ministry of Health. In the other Member States, they are part of the Agriculture administration, focusing mostly on economic aspects of farms animal production.

Main purpose of VPH is Prevention and veterinary Public Health activities are meant to safeguard and promote positive man-animal-environmental relationship. The main focus relates to security and safety of food derived from animals, but research, comparative pathology, animal assisted therapy are also relevant issues dealt by the veterinary service organization. Veterinary services have to deal also with the negative aspects that might ensue, from the sanitary point of view from the control of zoonotic agents to pest control (Donelli et al. 2004).

Activities related to VPH are considered an indicator of the state of peace and prosperity of a country, as can be inferred observing the difference existing between developing countries and developed countries in dealing with veterinary public health issues. Culture, religion, economic and social situation all play a relevant role in defining priorities and actions (Baldelli 2012).

Today the veterinary relevance in public health is widely recognized and international organizations, such as WHO, FAO and the 'World Organization for Animal Health (OIE) have emphasized the contribution of veterinary activities in Public Health, in the context of the "One Health" concept. But, it is important to take into account, that in Italy, Veterinary Public Health is the basis of the veterinary culture and organization since the inception of Italian state in the XIX century.

In the past, at international level, the concept of Public Health referred exclusively to the protection of human health, today the three organizations: FAO, WHO and OIE have defined an alliance to promote the concept of "one health" in the whole planet finally recognizing the fundamental contribution of veterinary science to the health of human beings.

The article focuses on the worldwide activities presently carried out to pursue global health, and on the importance of Italian veterinary public health concept in fostering sanitary levels not only in developed countries, but also in developing countries for a better world. In Italy this it is not a novel concept: it is the basis of the veterinary culture and of the activity of Veterinary Services since their foundation.

2   EXPO Event, Milano 2015

The theme that has been identified for Expo 2015, in Milan, was: " Feeding the Planet , Energy for Life", and if we read the "Theme Guide", there are many arguments that put this issue in the context of globalization, food safety, food security, issues on which the Italian VPH is focused historically and it constitutes its core .

The "Expo Milan 2015 Theme" is linked to certain objectives of the United Nations Millennium Development Goals, including:

– the first: to eradicate extreme poverty and hunger, and reduce by 50%, the proportion of people throughout the world who suffer from hunger;
– the fourth: to reduce by two thirds the mortality rate in children under the age of 5 years;
– the fifth: to improve maternal health and, in particular, reduce the maternal mortality rate by three quarters;
– the seventh: to ensure environmental sustainability, in particular by supporting sustainable development policies and programmes in order to reverse the loss of environmental resources and reduce biodiversity loss;
– the eighth: to develop a global partnership for Development.

The socio-cultural approach to the Theme embraces all scientific and educational programs aimed at educating people in healthy and balanced diets that must be implemented by countries, scholastic institutions, families, businesses and civil organizations. The purpose of these programmes is to harmonize respectful relations between individuals and their environment via fair and equal access to resources by the global population, without waste or unfairness, and to promote encounters and exchanges between people with different social and cultural identities and different food traditions, that are seen as a form of intangible cultural heritage linked to language, arts and traditional jobs and crafts.

The Guide take also into account the Cooperation for a Development Approach.

The Theme, 'Feeding the Planet, Energy for Life' can be described as methods and tools for cooperation having the purpose to reduce hunger, malnutrition and social imbalances linked to access to food, as well as any programme aimed at distributing investments in order to obtain effective results in developing poor rural areas and urban areas in developing countries.

Partnership agreements take special relevance here because they seek to respect the roles and specific needs of the people directly involved. (Theme Guide, EXPO Milan 2015).

All the aspects highlighted by EXPO Theme Guide, refer to: "the cycle of food, linked to food safety, food security regarding all people and all social status of the population".

The role of Veterinary Public Health is highlighted and International Cooperation must have the aim to share scientific and technical knowledge between countries.

3   FAO Contribution to Public Health

The economy of developing countries, depends on the availability of healthy animals for safe food. Animal Health ensures work capacity, food production and the access to the international market of animals and products of animal origin that are often a significant part of the Gross Domestic Product (GDP) being the economy of the Developing countries linked to agricultural production.

In some countries the economic survival of a family is dependent from the presence of few animals that are the only wealth and livelihood.

In the agricultural societies, with economy linked to agriculture, usually the poorest countries in the world, it is logical that the social and political stability is influenced by the ability to guarantee the people a minimum income and freedom from hunger.

It is scientifically demonstrates that the only remedy to war, especially in Africa, is to ensure

access to water and availability of food (Parodi et al. 2011).

Of great actuality and importance is the lecture of the FAO General Director – Graziano De Silva, in September 2015, stressing links between conflict, imperiled rural livelihoods and migration (FAO 2015).

The millions of people who are being forced to flee from war, poverty and other hardships are a tragic reminder of the urgent need for peaceful solutions based on social justice and improved economic opportunities for all. Key to achieving this is the protection of and investment in rural livelihoods.

> Rural development and food security are central to the global response to the refugee crisis. War causes hunger and hunger too, kills and forces people from their homes (FAO News 2015).

Migrations for environmental causes, including food crises, will become more frequent in the near future, assuming that the migration phenomenon is definitely caused to escape hunger. In a scenario where desertification, pollution and overheating of the atmosphere threaten the survival of the African people by pushing them to a biblical exodus towards Europe, causing concerns in the Old Continent (Giordano 2015).

There is no single solution to ensure food for the entire world population, the report "The State of Food Insecurity in the World 2015 – SOFI" (FAO 2015, Hunger Report) highlights several factors that can play a key role in achieving the goal of eliminating world hunger.

Inclusive growth provides opportunities for those with meagre assets and skills, and improves the livelihoods and incomes of the poor, especially in agriculture. It is therefore among the most effective tools for fighting hunger and food insecurity, and for attaining sustainable progress.

Enhancing the productivity of resources held by smallholder family farmers, fisher folk and forest communities, and promoting their rural economic integration through well-functioning markets, are essential elements of inclusive growth.

Social protection contributes directly to the reduction of hunger and malnutrition.

4 The Lancet Commission

The 2015 Lancet Commission on Health and Climate Change has been formed to map out the impacts of climate change, and the necessary policy responses, in order to ensure the highest attainable standards of health for populations worldwide. This Commission is multidisciplinary and international in nature, with strong collaboration between academic centers in Europe and China (Watts et al. 2015).

The central finding from the Commission's work is that tackling climate change could be the greatest global health opportunity of the twenty-first century. The key messages from the Commission are the environment, food safety and food security, water quality, air pollution, land use change, ecological change, in addition to mobility and conflict status.

It is clear that, at international level, the sources of animal origin food require the availability of qualified Veterinary Services, of well-trained biologists, and of human and material resources necessary to verify and ensure the health of animals and assistance to farmers, to reach the necessary sanitary conditions of food and also to improve the profitability of farms and agricultural production in general.

5 FAO/OIE/WHO "Tripartite Alliance – Concept Note"

In order to achieve more effective management of zoonotic and high impact diseases in the future, there is a global need to improve diagnostics, especially utilizing the new tools as genomics instruments, data analysis and risk assessment, epidemiology, social science and communication. Linking expert institutions through global networks within both the human and animal health sectors would enable new real-

time systems where methodology, data availability and responsibilities are shared both horizontally and vertically. Improved networking among countries promotes trust, transparency and cooperation. FAO, OIE and WHO are committed to working more closely together to align activities related to the human – animal – ecosystems interface in order to support Member States. The emergence of new or the re-emergence of existing animal diseases, including zoonoses, the growing threat of transboundary animal diseases, the impact of environmental changes and globalization, as well as new societal demands related to food security, food safety, public health and animal welfare, emphasize the critical need for collaboration between the three organizations.

While not a new concept, the OIE endorses the "One Health" approach as a collaborative and all-encompassing way to address, when relevant, animal and public health globally. This collaboration should not be limited to only the international level, but must be translated as a new and fundamental paradigm at national levels.

Promoting a collaborative "One Health" approach at national levels will result in a deeper and sustainable political support for the coordinated prevention of high public health and animal impact diseases at the human-animal interface.

Recent efforts in controlling emerging pandemic diseases and contributions towards pandemic preparedness have re-emphasized the need for enhanced collaboration to reduce risks of zoonotic potential including foodborne diseases and severe animal diseases at its source. Therefore, the OIE continues to insist on the critical need of constant improvement of the veterinary governance and its cooperation with public health managers and to consider all relevant activities as a global public good.

Within the framework of the Tripartite Alliance, WHO, FAO, and OIE recognize their respective responsibilities in fighting diseases, including zoonoses, that can have a serious health and economic impact. They have been working together for numerous years to prevent, detect, control and eliminate disease risks to humans originating directly or indirectly from animals. In 2010, the FAO/OIE/WHO Tripartite Concept Note (April 2010) officially recognized this close collaboration, with joint strategies at the human-animal-environment interface, to support their Member Countries. Three priority areas of work were defined: zoonotic influenzas, rabies and the fight against antimicrobial resistance. (Concept Note, April 2010; FAO OIE WHO Concept Note 2010).

6   The "One Health" Concept

Currently, countries are facing the occurrence of infectious diseases spreading from country to country, regardless of political borders, and this is a challenge for institutions and veterinary public health services that must identify new and more holistic control approaches (Seimenis 2008).

Health experts from around the world met on September 29, 2004, for a symposium focused on the current and potential movements of diseases among human, domestic animal, and wildlife populations, organized by the Wildlife Conservation Society and hosted by The Rockefeller University.

The product, called "The Manhattan Principles" by the organizers of the "One World, One Health" event, lists 12 recommendations for establishing a more holistic approach to preventing epidemic/epizootic diseases and for maintaining ecosystem integrity for the benefit of humans, their domesticated animals, and the foundational biodiversity that supports us all (Wildlife Conservation Society 2004).

It is important to note that, before, in 1964, C.W. Schwabe enunciated the concept of one medicine in Schwabe's' Veterinary Medicine and Human Health: a unique approach for human, animal and zoonosis (Schwabe 1964, 1984; Cardiff et al. 2008; Battelli and Mantovani 2011).

The concept in 2004 has gained the attention at international level and One Health stands for the health of humans, animals and the environment as one entity. The interdisciplinary

cooperation creates synergies that generate added value and is the promising strategy to strengthen health systems.

During the last two decades, great concern has been expressed on the emergence and re-emergence of certain infectious diseases, among them zoonotic diseases play an important role (Meisser et al. 2011), with an holistic approach recognizing that the human and others animals health is ONE and they share the same environment with reciprocal influence (Seimenis 2008).

This has focused attention on the concept of emerging – reemerging diseases, most of which are zoonotic diseases, that may cause a devastating impact on humans, animals and the environment. Their emergence is influenced by socio – economic, environmental and ecological factors and their interaction must be seen as a unique system in which the different components play a different role:

- the effects of globalization on trade of animals and animal products and the need for a multinational holistic approach;
- the role of wildlife and of environmental factors in the spread and maintenance of infections;
- the fundamental importance of the integration of veterinary and medical treatments into "one medicine" strategy and the necessity to pursue a multidisciplinary approach;
- the need for a comprehensive view over the whole production chain, following a "farm to fork" approach;
- migratory flows leading to the appearance in new territories of "old" pathogens and the discovery of "new" pathogens into newly discovered ecosystems;
- adaptation of pathogens into new hosts or into different ecological conditions;
- the ability of carriers to adapt to different environmental conditions in respect of the traditional ones.
- the importance to have diagnostic tools available to discover in "real time" the presence of microbial circulation in case of new pathogens in new territories (Lelli et al. 2011).

Food security is a complex topic, often among the causes of geopolitical crises and it is essential to build up a sustainable future. However, it implies insights, and coordinated and incisive interventions at European and International level. Numerous are the connections and interdependence between the food sector – in its various aspects -and the social, cultural, geopolitical and strategic implications. Various issues are examined: land grabbing, geopolitical power, land management, biosecurity, including, in specific cases in Africa and Asia, financial aspects and migration. All, without forgetting a new approach of intelligence in the food industry to identify the limits of an integrated threat capable of adding espionage, economic competition and geopolitical ambitions (Linzi 2015).

7 European Union and Public Health

Protecting the health of humans, animals and plants at every stage of the food production process is a public health and economic priority. The European Union's (EU's) food safety policy aims to ensure that EU citizens enjoy safe and nutritious food produced from healthy plants and animals, whilst enabling the food industry — Europe's largest manufacturing and employment sector — to operate in the best possible conditions.

EU policy safeguards health along the whole 'agro-food chain' — every part of the food production process from farming to consumption — by preventing food contamination and promoting food hygiene and information, plant health and animal health and welfare.

The basic principles for the EU's food safety policy are defined in the EU's General Food Law, adopted in 2002. Its general objectives are to facilitate the free trading of food across all EU Member States by ensuring the same high level of consumer protection in all Member States (Regulation EC No 178/ 2002 European Commission).

8 Italy: An Essential Element for Public Health

Public Health concept is essential to ensure health and welfare of human and animal population, to ensure their correct integration and interrelation. Since the beginning of the slogans: "One Health", "One Health-One Medicine", "One World-One Health", "One Medicine-One Science", this concept has been endorsed internationally, and it is notable that, in some countries, including Italy, the awareness that human health, animals and environment is a unique and ancient concept. In Italy since the Renaissance, and formally in the last 100 years, the Public Health System, including Veterinary Services, was organized in line with this vision (Battelli 2014).

Improving animal health and welfare increases availability and quality of proteins for the human population and preserve farmland abandonment, ensure food safety, prevents human health problems. Therefore, the Community has to ensure the function of the official Veterinary Services "from stable to table".

Italy, recognizing the role of the official veterinarian as UNIQUE, taking into account that health and well-being of both man and animals are interrelated, has institutionalized Community official veterinary services, since 1930.

As an example, a 1934 law obliged each Municipality:

- to build a slaughterhouse as a mandatory "public health infrastructure"
- to employ Communal veterinary officers
- to care for the health and welfare of animals in the farms
- to manage slaughterhouses
- to carry out food inspection in the Municipality food retailers (Italian Law *Regio Decreto* 1934 n. 1265; Caporale 2011).

At that time all the measures to modernize and strengthen the technical structures of veterinary services were taken, to update prophylactic action focused to modern principles and proven experience, expansion and intensification of operative action in the field of the restoration of animal diseases dangerous for human beings, the availability of more substantial financial resources to prevent or to eradicate some serious animal diseases.

This approach led rapidly to recognize, already in the middle of the XX Century that health of humans and other animals was ONE and that it was more effective and efficient to prevent diseases in animal populations, in order to protect human health, and the role of Veterinary Public Health in the field of food safety, food security and prevention of human diseases in humans, has established that the Directorate General of Veterinary Services, since its establishment, in 1946, was part of the Ministry of Health in Italy (Italian Law DPR n. 264 1961; Italian Law n. 101 1974).

In conclusion, to work at the global level, to ensure ONE Health, there is the need to developed the culture of:

– inclusion than exclusion and supremacy
– sharing of knowledge with respect to the ownership of knowledge
– organization to share knowledge than hierarchical organization

The exchange of data, field material, biological samples, diagnostic reagents and reference material allows the growth of staff skills and the scientific level of laboratories both in Italy and in countries with which collaboration and cooperation is in place.

New laboratory methods and new vaccines, by biotechnology instruments, for the prevention of diseases in animals and humans have to be developed and validated, interlaboratory tests have to be organized to assess the competences of the personnel (Caporale 2014; Caporale et al. 2009).

The contribution of the Veterinary Services to public health and poverty reduction in developing countries is increasingly a core value, to provide an appropriate contribution, as well as to ensure a sustainable future when the human population will reach the milestone of 9 billion people on Earth.

A review of University *curricula* to include 'One Health' precepts within a livelihoods framework should be an urgent undertaking, if veterinarians are to remain relevant to the context in which they live and work, with a holistic approach with regard to health of humans, animals and environment (Muma et al. 2014).

This should also include competence in analyzing the costs and benefits of the control strategies they institute, to assure the best response to the needs of the society. Many disease control campaigns have been conducted without a full assessment of their economic impacts both in terms of the cost of conducting the exercise, disposal, and the effects of other related non-agriculture industries such as tourism.

It will be necessary to ensure the level of food hygiene, improve production, the defense of rural farming and rural traditional productions, as they are very popular in the territories that need a contribution to the development of the economy of the Countries.

References

Baldelli R (2012) Sanità Pubblica Veterinaria: dove? Available from: https://www.google.it/url?sa=t&rct=j&q=&esrc=s&source=web&cd=2&cad=rja&uact=8&ved=0ahUKEwi93bCD_djTAhXBvhQKHdOfAf4QFggtMAE&url=http%3A%2F%2Fwww2.vet.unibo.it%2Fstaff%2Fbaldelli%2FSanit%25C3%25A0%2520Pubblica%2520Veterinaria%25202008.2009%2FPresentazioni%2F2.%2520concetto%2520di%2520SANITA%27%2520PUBBLICA%2520VETERINARIA.pdf&usg=AFQjCNEXmyxEiCq50V1Zfd3jlqhIZsKbCA&sig2=FdfJHgxiR4Sypaw0IjD4Mw. Last visited 5 May 2017
Battelli G (2014) Medicina unica – salute unica: per l'uomo, per gli animali e per l'ambiente. Giornale Italiano di Medicina Tropicale 2014 19(4):17–23
Battelli G, Mantovani A (2011) The veterinary profession and one medicine: some considerations, with particular reference to Italy. Vet. Ital. 47(4):389–395
Caporale V (2011) Sustainability of fight against infectious diseases of animals. Twinning projects in Africa, a bridge across countries to build a transboundary network. Journal of Commonwealth Veterinary Association – Proceedings of the 5th Pan Commonwealth Veterinary Conference – Special Issue July 2011; 27 (2)

Caporale V (2014) La funzione degli Istituti Zooprofilattici Sperimentali dal XIX al XXI secolo, una storia italiana di cultura, passioni, grandezza e varia umanità, pagg. 65–98 – Atti della Conferenza "La Medicina Veterinaria Unitaria (1861–2011)", 22 Giugno 2011, Ministero della Salute – Roma, Edito a cura della Fondazione Iniziative Zooprofilattiche e zootecniche, Brescia- ISBN 978–88–97562-08-5: 65–98

Caporale V, Lelli R, Scacchia M, Pini A (2009) Namibia: un esempio di cooperazione internazionale per lo studio delle patologie emergenti. Vet Ital 45(2):243–248

Cardiff RD, Ward JM, Barthold SW (2008) 'One medicine – one pathology': are veterinary and human pathology prepared? Lab Investig 88:18–26

De Silva G (2015) Food and Agriculture Organization. FAO Director-General: food security and the migration crisis – Available from: http://www.fao.org/news/story/it/item/332007/icode/. Last visited 16/02/2016

Donelli G, Lasagna E, Macrì A, Mantovani A (2004) Sull'afferanza dei Servizi Veterinari all'amministrazione Pubblica Italiana: una ricostruzione storica. 35th International Congress of the World Association for the History of Veterinary Medicine Grugliasco (Turin), Italy, September 8–11, 2004

European Regulation (EC) No 178/2002 of the European Parliament and of the Council of 28 January 2002 laying down the general principles and requirements of food law, establishing the European Food Safety Authority and laying down procedures in matters of food safety. Available from: http://europa.eu/pol/pdf/flipbook/it/food_it.pdf. Last visited 16 Feb 2016

EXPO, Milan 2015, Theme Guide. A cura del gruppo di lavoro TEG Tema Coordinamento editoriale Direzione Thematic Spaces. Con il contributo di Marco Amato e Stefano Karadjov. Render a cura di Centro Sviluppo Realtà Virtuale S.r.l. Available from: http://www.expo2015.org/it/cos-e/il-tema. Last visited 16 Feb 2016

Food and Agriculture Organization – World Animal Health Organization – World Health Organization (2010) The FAO-OIE-WHO Collaboration Sharing responsibilities and coordinating global activities to address health risks at the animal-human-ecosystems interfaces, A Tripartite Concept Note, April 2010. Available from: http://www.who.int/influenza/resources/documents/tripartite_concept_note_hanoi_042011_en.pdf. Last visited 16 Feb 2016

Food and Agriculture Organization (2015) Meeting the 2015 international hunger targets: taking stock of uneven progress. Hunger Report (The State of Food Insecurity in the World 2015 – SOFI). Available from: www.fao. org/hunger/en/. Last visited 16 Feb 2016

Giordano A (2015) Crisi alimentari, migrazioni e sicurezza. Gnosis rivista italiana di intelligence 2015; 1:115–121

Italian Law: Regio Decreto 27 luglio 1934, n. 1265.Testo unico delle leggi sanitarie. GU n.186 del 9–8-1934 – Suppl. Ordinario n. 186

Italian Law: Decreto Presidente della Repubblica (D.P.R.) 11 Febbraio 1961, n. 264. Disciplina dei servizi e degli organi che esercitano la propria attività nel campo dell'igiene e della sanità pubblica. G.U n. 100, 22 Aprile 1961

Italian Law: Legge 11 marzo 1974, n. 101. Modifica della legge 23 giugno 1970, n. 503, sull'ordinamento degli Istituti zooprofilattici sperimentali. GU n.100, 17 Aprile 1974

Lelli R, Monaco F, Cito P, Calistri P (2011) Factors influencing the risks of emerging animal diseases in the Mediterranean regions: a review. Giornale Italiano di Medicina Tropicale 2011 16(3–4):83–91

Linzi G (2015) Editoriale 1/2015. Gnosis rivista italiana di intelligence 2015; 1. Available from: http://gnosis.aisi.gov.it/Gnosis/Rivista42.nsf/servnavig/11. Last visited 16 Feb 2016

Meisser A, Schelling E, Zinsstag J (2011). One Health in Switzerland: a visionary concept at a crossroads? Swiss Med Wkly 141:w13201

Muma JB, Mwacalimba KK, Munang'andu HM, Matope G, Jenkins A, Siamudaala V, Mweene AS, Marcotty T (2014) The contribution of veterinary medicine to public health and poverty reduction in developing countries. Vet Ital 50(2):117–129

Parodi P, Venturi L, Flamini B, Mc Grath P (2011) International trade and public health in Africa: the example of Swaziland. Giornale Italiano di Medicina Tropicale 2011; 16(3–4):13–22

Schwabe CW (1964) Veterinary medicine and human health, vol xvii. Williams & Wilkins, Baltimore, pp 1–516

Schwabe CW (1984) Veterinary medicine and human health, vol xix, 3rd edn. Williams & Wilkins, Baltimore, pp 1–680

Seimenis A (2008) Zoonotic diseases in the Mediterranean region: a brief introduction. Vet Ital 44 (4):573–576

Sikkema R, Koopmans M (2016) One health training and research activities in Western Europe. Infect Ecol Epidemiol 6:33703. http://dx.doi.org/10.3402/iee.v6.33703

Travis DA, Sriramarao P, Cardona C, Steer CJ, Kennedy S, Sreevatsan S, Murtaugh MP (2014) One medicine one science: a framework for exploring challenges at the intersection of animals, humans, and the environment. Ann N Y Acad Sci 1334:26–44

United Nations. Economic and Social Affairs Population Division. 2014. World Population Prospects: The 2015 Revision Key Findings and Advanced Tables: ESA/P/WP.241

Watts N, Adger WN, Agnolucci P, Blackstock J, Byass P, Cai W, Chaytor S, Colbourn T, Collins M, Cooper A, Cox PM, Depledge J, Drummond P, Ekins P, Galaz V, Grace D; Graham H, Grubb M, Haines A, Hamilton I, Hunter A, Jiang X, Li M, Kelman I, Liang L, Lott M, Lowe R, Luo Y, Mace G, Maslin M, Nilsson M, Oreszczyn T, Pye S, Quinn T, Svensdotter M, Venesky S, Warner K, Xu B, Yang J, Yin Y, Yu C,

Zhang Q, Gong P, Montgomery H, Costello A (2015) Health and climate change: policy responses to protect public health. The Lancet Commission. www.thelancet.com. 23 June 2015. Available from: http://dx.doi.org/10.1016/S0140-6736(15)60854-6. Last visited 16 Feb 2016

Wildlife Conservation Society (2004) Summary of Conference One World-One Health: Building Interdisciplinary Bridges to Health in a Globalized World. New York, 29 September 2004. The Rockefeller University, New York Available from: http://oneworldonehealth.org/sept2004/owoh_sept04.html. Accessed 16 Feb 2016

World Health Organization. Future trends in veterinary public health. Gruppo di lavoro OMS: TE, Italy, 1999.

Available from: http://www.who.int/zoonoses/vph/en/. Last visited 16 Feb 2016

Zinsstag J (2013) Convergence of EcoHealth and One Health. EcoHealth (February 2013). doi:10.1007/s10393-013-0812-z

Zinsstag J (2016) Pathways to a sustainable world – Animal, human and environmental health. In: The 4th International One Health Congress & 6th Biennial Congress of the International Association for Ecology & Health. 3–7 December 2016, Melbourne, Australia

Zinsstag J, Schelling E, Waltner-Toews D, Whittaker M, Tanner M (2015) One health: the theory and practice of integrated health approaches, 1st edn. CABI, Oxfordshire

Adv Exp Med Biol - Advances in Microbiology, Infectious Diseases and Public Health (2018) 9: 73–83
DOI 10.1007/5584_2017_70
© Springer International Publishing AG 2017
Published online: 22 June 2017

Biofilm-Forming Ability and Clonality in *Acinetobacter baumannii* Strains Isolated from Urine Samples and Urinary Catheters in Different European Hospitals

Claudia Vuotto, Filipa Grosso, Francesca Longo,
Maria Pia Balice, Mariana Carvalho de Barros, Luisa Peixe,
and Gianfranco Donelli

Abstract

Objective

Biofilm formation has been associated with the persistence of *Acinetobacter baumannii* in hospital settings and its propensity to cause infection. We investigated the adhesion ability and clonality of 128 A. baumannii isolates recovered from urine and urinary catheters of patients admitted to 5 European hospitals during 1991–2013.

Methods

Isolates identification was confirmed by rpoB sequencing and by the presence of blaOXA-51. The presence of carbapenemases was detected by PCR. Clonality was determined by Sequence Group (SG) identification, Pulsed field gel electrophoresis (PFGE) and Multilocus sequence typing. Adhesion ability was defined by quantitative biofilm production assay and biofilms were characterized by Confocal Laser Microscopy and Scanning Electron Microscopy.

Results

The 128 isolates, either resistant (85.9%) or susceptible (14.1%) to carbapenems, and belonging to 50 different PFGE types and 24 different STs, were distributed among SG1 (67.2%), SG2 (10.2%) and other allelic profiles (22.7%). ST218 was the most frequent ST, corresponding to 54,5% of the isolates collected between 2011 and 2013.

C. Vuotto (✉), F. Longo, and G. Donelli
Microbial Biofilm Laboratory, IRCCS Fondazione Santa Lucia, Rome, Italy
e-mail: c.vuotto@hsantalucia.it; f-longo@hotmail.it; g.donelli@hsantalucia.it

F. Grosso and L. Peixe
REQUIMTE. Laboratório de Microbiologia, Faculdade de Farmácia, Universidade do Porto, Porto, Portugal
e-mail: filipagrosso@gmail.com; lpeixe@ff.up.pt

M.P. Balice
Clinical Microbiology Laboratory, IRCCS Fondazione Santa Lucia, Rome, Italy
e-mail: mp.balice@hsantalucia.it

M.C. de Barros
Microbial Biofilm Laboratory, IRCCS Fondazione Santa Lucia, Rome, Italy

REQUIMTE. Laboratório de Microbiologia, Faculdade de Farmácia, Universidade do Porto, Porto, Portugal
e-mail: marianabarros412@gmail.com

Among the 109 isolates showing resistance to at least 1 carbapenem, 55% revealed the presence of an acquired carbapenem-hydrolyzing class D - lactamases (CHDL): blaOXA-23 were the most frequent gene detected from 2008 onwards (75%). Among all the clinical isolates, 42.2% were strong biofilm producers, with the older isolates having the highest adhesion ability. Most isolates recovered later, belonging to ST218 and harbouring blaOXA-23, were homogeneously less adhesive.

Conclusions

An evolution towards a decrease in adhesion ability and a CHDL content change was observed along the years in several European countries.

Keywords

Acinetobacter baumannii · Clonality · Biofilm · Urinary infection · Carbapenem resistance

1   Introduction

Acinetobacter baumannii is one of the emerging pathogens frequently associated with hospital-acquired infections (Antunes et al. 2014). This species is able to cause, particularly in compromised patients, a range of serious infections, such as bacteremia, pneumonia, meningitis, wound infections and urinary tract infections (McConnell et al. 2013; Dettori et al. 2014; Gniadek et al. 2016). Biofilm formation is an important feature of most clinical isolates of *A. baumannii* and it is known that biofilm producers very effectively colonize biotic as well as medical devices surfaces, such as indwelling urinary catheters (Djeribi et al. 2012; Longo et al. 2014; Zarrilli. 2016). This characteristic increases the survival of *A. baumannii* in clinical settings and contributes to its persistence in the hospital environment, thus increasing the probability of nosocomial infections (Espiral et al. 2014).

Carbapenems, introduced since 1985, have been for many years the most effective antibiotics for the treatment of MDR *A. baumannii* infections, but a worldwide increasing frequency of *A. baumannii* isolates resistant to carbapenems has been recognised in the last decade (Poirel and Nordmann. 2006; Peleg et al. 2008; Jones et al. 2014; Gniadek et al. 2016). In fact, a number of outbreaks of carbapenem-resistant *A. baumannii* have been reported not only in Europe, North and Latin America, but also in Australia and other countries including Tunisia, South Africa, China, Taiwan, Singapore, Hong Kong, Japan, South Korea, and the remote French Polynesia (Gniadek et al. 2016).

This emerging issue worldwide highlights the need to characterize their clonality and mechanisms of antimicrobial resistance as well as to investigate other features that could be involved in their resilience in the hospital setting.

Early genotypic-based analysis delineated two major groups among *A. baumannii* isolates, with the pan-European clonal lineage II [which corresponds to Sequence Group (SG) 1] being widely disseminated (Adams-Haduch et al. 2011; Grosso et al. 2011), followed by clonal lineage I (SG2 and SG5) (Zarrilli et al. 2013). Based on a Multi Locus Sequence Typing (MLST) analysis, it was possible to further distinguish among different clones within each clonal lineage, some of them widespread (e.g. ST92 of SG1 and ST109 of SG2) (Zarrilli et al. 2013). Acquisition of carbapenemase genes, mostly carbapenem-hydrolysing class D β-lactamases (CHDL), has been identified as the main mechanism of carbapenem resistance, with the siderophore TonB-dependent receptor often described in plasmids carrying the CHDL bla_{OXA-40} (Grosso et al. 2012; Mosqueda et al. 2014). The correlation between antibiotic resistance profile and biofilm formation in general (Qi et al. 2016), and between *A. baumannii* particular clones and biofilm production in particular, has been barely explored (Eijkelkamp et al. 2011; Rao et al. 2008; Rodríguez-Baño et al. 2008). Moreover, very few studies assessed a possible correlation between production of

CHDLs and biofilm-forming ability (Luo et al. 2015).

The aim of this study was to gain insight on adherence, biofilm-forming ability and clonality in a comprehensive collection of *A. baumannii* clinical isolates obtained from urine samples and urinary catheters of patients admitted to different European hospitals.

2 Materials and Methods

A total of 128 *Acinetobacter baumannii* clinical isolates, randomly collected during the period 1991–2013 from urine samples (n = 75) and urinary catheters (n = 53) of patients hospitalized in different European clinical settings of Croatia (n. 9; 1 hospital), Czech Republic (n. 8; 1 hospital), Germany (n. 12; 2 hospitals), Italy (n. 71; 5 hospitals), and Portugal (n. 28; 3 hospitals), were included in this study. The reference strain *A. baumannii* ATCC 19606 was used as positive control for biofilm production assays. The isolates were initially identified by different diagnostic systems, depending on the hospital source, including Vitek2, API 20 NE, Phoenix BD, and MALDI-TOF MS. Species identification was confirmed with bla_{OXA-51} detection (Woodford et al. 2006), and *rpoB* partial gene amplification and sequencing (Gundi et al. 2009).

Antimicrobial susceptibility testing for carbapenems was determined by different automated systems and disc diffusion method according to CLSI guidelines (CLSI 2011). The acquired CHDLs were detected by PCR (Woodford et al. 2006).

Clonal relatedness among isolates was established by *Apa*I -pulsed-field gel electrophoresis (PFGE), PCR identification of *A. baumannii* Sequence Groups (Turton et al. 2007) and MLST (Bartual et al. 2005). PFGE patterns were analysed by InfoQuestTM FP software version 5.4 (BioRad Laboratories) with Dice's coefficient analysis of peak positions executed. The unweighted-pair group method using average linkages (UPGMA) was applied, and the bandwidth tolerance was set at 1.5%. Isolates clustering together with >85% level of similarity were considered to belong to the same PFGE type (Seifert et al. 2005). Allele, Sequence Type (ST) assignments, new allelic sequences and/or profiles submissions were performed at *A. baumannii* MLST webpage, according to the Oxford scheme (http://pubmlst.org/abaumannii/).

Biofilm-forming ability was assessed by using a semi-quantitative biofilm production assay and each test was performed in triplicate and repeated at least 3 times. Briefly, for each isolate, 3 wells of a 96-well flat-bottomed plastic tissue culture plate were filled with 180 µl of LB supplemented with 1% glucose and 20 µl of the overnight culture diluted to a final optical density 600 (OD_{600}) = 0.1. After incubation at 37 °C for 24 h, each well was washed three times with PBS, dried for 1 h at 60 °C and stained for 15 min with 180 µl of 2% Hucker's crystal violet. The dye bound to the adherent cells was solubilised with 180 µl of 33% (v/v) glacial acetic acid. The OD of each well was measured at 570 nm by using a spectrophotometer (Multiscan FC; Thermo Scientific). The cut-off OD (ODc) was defined as three standard deviations above the mean OD of the negative control. According to the defined ODc, all the isolates were classified on the basis of their ability to adhere into the following categories: non-adherent (OD ≤ ODc), weakly adherent (ODc < OD ≤ 2xODc), moderately adherent (2xODc < OD ≤ 4xODc), and strongly adherent (4xODc < OD) (Donelli et al. 2012).

Imaging by field emission scanning electron microscopy (FESEM) and confocal laser scanning microscopy (CLSM) was performed for three representative isolates of the most highly adherent isolates. Briefly, for FESEM, each well of a 24-well plastic tissue culture plate, with a 10-mm diameter glass coverslip placed on the bottom, was filled with 200 µl of a broth culture adjusted to OD_{600} = 0.1 and 1.8 ml of LB broth supplemented with 1% glucose and incubated for 24 h at 37 °C. Biofilms were washed carefully three times with PBS, then fixed with 2.5% glutaraldehyde in 0.1 M cacodylate buffer (pH 7.4) at room temperature for 30 min and finally washed with the same buffer. Thus, samples were post-fixed with 1% $OsO4$ in 0.1 M phosphate buffer for 20 min, washed again with 0.1 M cacodylate buffer and dehydrated through graded ethanol solutions (30%, 50%, 70%, 85%, 95%, 100%). After critical point drying by K850 critical point dryer (Quorum Technologies) and gold coating by Q150R S Rotary-Pumped Sputter Coater (Quorum Technologies), biofilm samples were examined by FESEM (Sigma-Zeiss) at an accelerating voltage of 5 kV (Vuotto and Donelli 2014). For CLSM, biofilms were formed as described above and fixed with 3.7% paraformaldehyde for 30 min. Then, biofilms were washed with distilled water and stained for 15 min with the SYTO® 9 green fluorescent nucleic acid stain (Invitrogen, Molecular Probes®) by adding, in each well 1.5 µl of the dye in 1 ml of distilled water for 15 min at room temperature in the dark. The stain was aspirated, and the coverslips gently washed twice with distilled water. The SYTO® 9 stain labels both live and dead bacteria that fluoresce green (480/500 nm). Fluorescence was viewed by a CLSM (Nikon C1si).

Statistical analyses were performed using two-tailed, unpaired Student's t-test with GraphPad Prism software (GraphPad Software Inc., La Jolla, CA, USA). Results were expressed as means ± standard deviations (s.d.) of three independent experiments. Differences were considered as statistically significant when P-value was <0.05.

3   Results and Discussion

This study included the characterization of diverse STs, grouped in SG1 (7 STs, 24 PFGE types, n = 86; 2001–2013), SG2 (3 STs, 8 PFGE types, n = 13; 1991–2013) or non-SG1-non-SG2 lineages, [14 STs (8 new STs), 20 PFGE types, n = 29; 1991–2013], with SG1 dominating in recent years (Table 1), which is consistent with *A. baumannii* clonal evolution described worldwide (Zarrilli et al. 2013). German isolates presented higher clonal diversity, with 12 isolates

(n = 4 in 1991, n = 1 in 2009, n = 1 in 2010, and n = 5 in 2012) distributed among 11 STs, mainly from lineages other than SG1 (Table 1). On the contrary, in Italy and Portugal, ST218, from SG1, was the most frequent ST and corresponded to 54 out of 99 isolates (54.5%) collected in fairly recent years (2011–2013) (Table 1).

Most (85.2%) of the analysed *A. baumannii* isolates were resistant to carbapenems, mainly from Portugal and Italy, while the other countries showed variable percentages of carbapenem resistance (58.3% in Germany, 50% in Czech Republic and 11.1% in Croatia). These differences might be related to the collection time; Italian and Portuguese strains have been isolated in relatively recent years (2011–2013), while most of the isolates from Germany Croatia, and Czech Republic were back far enough in time (1991; 2001–2007; 2007–2008; respectively). Acquired CHDL were present in 55% of the carbapenem-resistant isolates: 75% displayed bla_{OXA-23}, 11.7% bla_{OXA-58}, 11.7% $bla_{OXA-24/40}$ and 1.7% bla_{OXA-72} (Table 1). Furthermore, bla_{OXA-23} was the only gene detected from 2008 onwards, with a couple of exceptions only for a Croatian isolate (bla_{OXA-72}) in 2010 and an Italian isolate ($bla_{OXA-24/40}$) in 2013 (Table 1).

The semi-quantitative biofilm production assay revealed that 42.2% of the isolates exhibited strong adherence properties, with ODs > 0.6 (strong biofilm producers' cut-off), while 10.9% of isolates were moderately adherent, 28.1% weakly adherent and 18.8% non-adherent. The strongly adherent strains were found in 60.4% of those isolated from urinary catheters (n = 53) and in 29.3% from urines (n = 75). Unexpectedly, strongly adherent isolates from urine samples were more frequently found among SG1 isolates, while among the other lineages they were mainly collected from urinary catheters. Isolates from SG2 and non-SG1-non-SG2 lineages, frequently demonstrated a strong adherence phenotype (69.2% and 79.3%, respectively), particularly those belonging to ST103 (OXA-58 producers) (Fig. 1b; Table 1). SG1 included the majority of isolates with acquired CHDLs (Table 1), those

Table 1 Characteristics of the 128 *A. baumannii* urinary strains isolated in different European countries

Lineages	ST (n)	Years	PFGE types	Countries	Origin[b] (n)	Carbapenem resistance (n/n_{tot})[c]	Acquired CHDL (n)	Highly adherent isolates (%)[d]
SG1 (European Clone II)	ST98 (5)	2003–2007	K	Portugal	U (2), UC (3)	5/5	OXA-24/ 40 (5)	U (100%), UC (100%)
	ST208 (1)	2008	K3	Portugal	UC (1)	1/1	OXA-23 (1)	UC (100%)
	ST195 (1)	2010	B	Germany	UC (1)	1/1	OXA-23 (1)	UC (0%)
	ST348 (1)	2010	J	Croatia	UC (1)	1/1	OXA-72 (1)	UC (0%)
	ST218 (55)	2010–2013	X, X1-X4, BB, BB1, BB2, CC, A, K5	Germany, Italy, Portugal	U (44), UC (11)	55/55	OXA-23 (33)	U (22,73%), UC (18,18%)
	ST451 (15)	2011–2013	AA, AA2, AA3, DD, DD1	Italy	U (11), UC (4)	15/15	OXA-23 (4)	U (45,45%), UC (25%)
	ST281 (8)	2012–2013	Y, JJ, CC, V, JJ1	Italy	U (8)	8/8	OXA-23 (4)	U (12,5%)
SG2 (European Clone I)	ST16 (1)	1991	E	Germany	UC (1)	0/1		UC (100%)
	ST439 (1)	1991	G	Germany	UC (1)	0/1		UC (100%)
	ST441 (11)	2001–2009	C,H, H2, EE, EE2, EE3	Croatia, Czech Republic, Germany	UC (11)	3/11	OXA-24/ 40 (1)	UC (63,64%)
Other allelic profiles	ST15 (1)	1991	F	Germany	UC (1)	0/1		UC (100%)
	ST514 (1)	1991	D	Germany	UC (1)	0/1		UC (100%)
	ST103 (7)	2001–2004	M	Portugal	U (3), UC (4)	7/7	OXA-58 (7)	U (33,33%), UC (100%)
	ST740 (1)[a]	2001–2007	H1	Croatia	UC (1)	0/1		UC (100%)
	ST775 (1)[a]	2001–2007	H2	Croatia	UC (1)	0/1		UC (100%)
	ST236 (2)	2012	M, N	Italy	U (2)	1/1	OXA-23 (1)	U (100%)
	ST391 (1)	2012	Q	Germany	UC (1)	1/1		UC (0%)
	ST732 (2)[a]	2012	R	Germany	U (1), UC (1)	2/2		U (100%), UC (100%)
	ST733 (1)[a]	2012	U	Germany	UC (1)	1/1	OXA-23 (1)	UC (0%)
	ST776 (2)[a]	2012	L, L1	Italy	UC (2)	1/2		UC (100%)

(continued)

Table 1 (continued)

Lineages	ST (n)	Years	PFGE types	Countries	Origin[b] (n)	Carbapenem resistance (n/n_{tot})[c]	Acquired CHDL (n)	Highly adherent isolates (%)[d]
	ST231 (4)		Z, Z2, HH, T	Germany, Italy	U (2), UC (2)	3/4	OXA-24/ 40 (1); OXA-23 (1)	U (50%), UC (100%)
	ST1159 (4)[a]	2007–2008	DD, FF, GG	Czech Republic	U (1), UC (3)	2/4		U (100%), UC (100%)
	ST1160 (1)[a]	2013	II	Italy	UC (1)	0/1		UC (0%)
	ST1164 (1)[a]	2013	KK	Italy	U (1)	1/1		U (100%)

[a]new STs described in this study
[b]urine (U) or urinary catheter (UC)
[c]isolates resulted to be resistant to carbapenems by antimicrobial susceptibility testing
[d]these values stand for the percentage of the highly adherent isolates out of the total number of isolates from U or UC in that particular ST

collected during the last period (ST195, ST218, ST451 and ST281; OXA-23) (Fig.1; Table 1) being significantly ($p \leq 0.001$) less able to form biofilm, when compared with the older ones, especially belonging to ST98 (OXA-24/40 producers), a long standing clone among Portuguese hospital settings (2001–2007) (Table 1). It seems that the dominant lineages nowadays are less able to adhere and form biofilm, with 52% of all the isolates collected from 1991 to 2011 demonstrating an OD > 2 while 60% of isolates collected during 2012–2013 showed an OD ≤ 1 and none of them exceeded OD = 2 (Table 1). Moreover, this behaviour seems to be accompanied with a CHDL content change (from OXA-24/40 and OXA-58 to OXA-23) together with an extension of antimicrobial resistance profile, confirming the worldwide trend on CHDLs type carried by *A. baumannii* isolates (Kohlenberg et al. 2009; Fu et al. 2010; Principe et al. 2014; Liu et al. 2015; Wu et al. 2015; Rolain et al. 2016; Viana et al. 2016). Interestingly, in a rabbit infection model it has been demonstrated that genes bestowing both intrinsic and acquired antibiotic resistance provided a positive *in vivo* fitness advantage to *A. baumannii* (Roux et al. 2015), as an additional feature explaining the success of the recent clones, even with decreased biofilm-forming ability.

On the contrary, Farshadzadeh and co-workers reported a significant association between strong biofilm formation and antimicrobial resistance (P = 0.0001) among bla_{OXA-23}-producing *A. baumannii* strains collected in Tehran from January 2012 to May 2013 (Farshadzadeh et al. 2015). However, these strains belonged to STs different from those analysed in our study.

In fact, very few studies on the possible correlations between *A. baumannii* clonality and biofilm production have been conducted so far, aiming to understand the mechanisms behind the global propagation of successful *A. baumannii* lineages. Sahl and colleagues sequenced seven genomes from the ST25 lineage and observed differences in biofilm formation between ST25 isolates and ST1, with the first producing a significantly more extensive biofilm (Sahl et al. 2015). In 2013, Giannouli and colleagues investigated some virulence-related traits of epidemic *A. baumannii* strains assigned to distinct STs, observing that the ability to form biofilm and to adhere to A549 pneumocytes was significantly higher for strains assigned to ST2

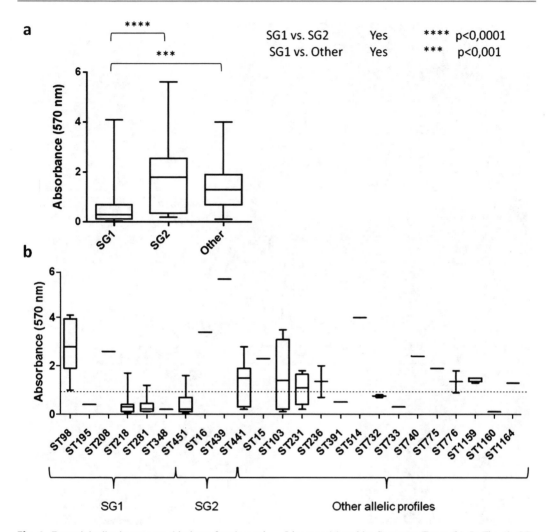

Fig. 1 Bacterial adhesiveness on abiotic surface by various groupings of the population of 128 *A.baumannii* clinical isolates collected from urine samples and urinary catheters in different European hospitals. OD_{570} values of *A. baumannii* population have been grouped by Lineages (**a**) and by Sequence Types (**b**). In Fig. 1b, STs have been listed according to the respective lineage and the dotted line indicates the cut-off for strong adhesion classification ($4 \times ODc = 0.6$). Statistical significance is included in the upper right corner and refers to Fig. 1a

(SG1), ST25 and ST78, in comparison with other STs (Giannouli et al. 2013). Although using different collections and methodologies, the above mentioned literature data, along with our results, seem to support the hypothesis that the distinct adhesion phenotypes observed are most probably due to differences in the lineages. The strong point of our analysis is represented by a larger population of isolates obtained from different European countries, thus providing a greater statistical power to our study. The collection of

A. baumannii isolates performed over the years provides also information on the variation in time of the CHDL content and on the gained or lost ability to form biofilm.

Finally, to analyse the ultrastructure and the three-dimensional organization of mature biofilms, 3 strongly adherent isolates belonging to the sequence types ST98 [HGSA25 ($ODs = 4.1 \pm 0.28$), Ac111 ($ODs = 3.8 \pm 0.28$)] and ST776 [AB1FSL ($ODs = 1.8 \pm 0.5$)] collected from urinary catheters, were selected for

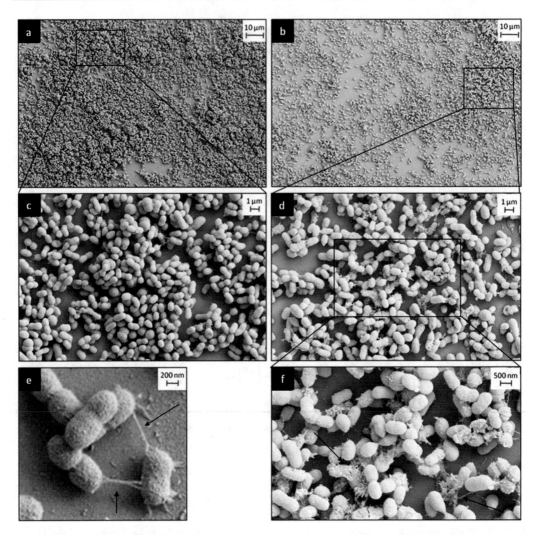

Fig. 2 FESEM micrographs of the biofilm-forming HGSA25 and Ac111 strains after 24 h incubation. HGSA25 biofilms observed at a magnification of 2,000X (**a**) and 10,000X (**c**) using an EHT = 5.00 kV; AC92 biofilms observed at a magnification of 2,000X (**b**) and 10,000X (**d**) using an EHT = 5.00 kV; (**f**) high resolution micrographs showing extracellular matrix observed at a magnification of 20,000X, (**e**) high resolution micrographs showing hair-like appendages observed at a magnification of 60,000X

further characterization by FESEM and CLSM in order to compare the biofilm ultrastructure among different lineages (Figs. 2 and 3). FESEM analysis of the ST98 isolates revealed the typical structure of *A. baumannii* biofilm characterized by water channels (Fig. 2), extracellular matrix (Fig. 2f), and pili (Fig. 2e). CLSM analysis of isolate belonging to ST776, revealed a sparse biofilm constituted by large bacterial clusters instead of the rich and homogeneous biofilm developed by the ST98 isolates (Fig. 3). It is tempting to speculate that this feature could have contributed to the persistence of ST98 in Portuguese hospitals for more than 15 years (Grosso et al. 2011). Thus, although these isolates were capable to form biofilm, the resulting structures were quite diverse, probably reflecting different mechanisms responsible for the sessile mode of growth which can also play a role for the different success of these lineages.

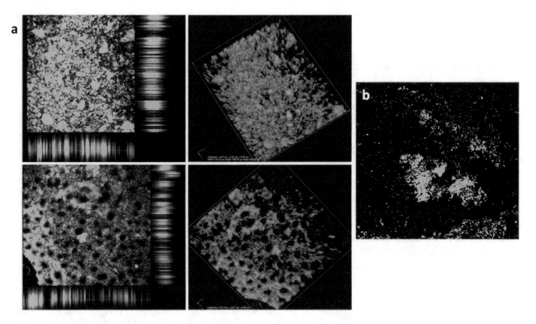

Fig. 3 CLSM images of the mature biofilms of three strongly adherent isolates belonging to ST98 [HGSA25 (ODs = 4.1 ± 0.3), and Ac111 (ODs = 3.8 ± 0.3)] and ST776 [AB1FSL (ODs = 1.8 ± 0.5)] collected from urinary catheters. In Fig. 3a are shown the images of horizontal (xy, large panel) and vertical (xz and yz, side panels) projections of HGSA 25 (image in the upper left) and Ac111 (image below on the left); three-dimensional images (Z-stacks) of biofilms produced by HGSA25 (image top right) and Ac111 (bottom right image) after 24 h incubation. In Fig. 3b bacterial clusters formed by AB1FSL isolate after 24 h incubation are displayed

4   Conclusion

We analysed a collection of clonally diverse *A. baumannii* isolates recovered from both urine samples and urinary catheters in different European hospital settings in the years 1991–2013, revealing a switch on dominant lineages that now seem to be less able to adhere and form biofilm, along with a CHDL content change (from OXA-24/40 and OXA-58 to OXA-23). Our data also suggest that the success of ST218, a recent widespread lineage, does not rely on its ability to adhere. Moreover, using FESEM analysis it was possible to reveal diverse biofilm structures that could account for the different resilience of *A. baumannii* isolates, namely in the hospital context.

Acknowledgments We thank (in alphabetical order) Edoardo Carretto (Clinical Microbiology Laboratory, IRCCS - Arcispedale Santa Maria Nuova, Reggio Emilia, Italy), Anna Giammanco (Department of Sciences for Health Promotion and Mother-Child Care "G. D'Alessandro", Palermo, Italy), Ivana Goic-Barisic (Clinical Department of Microbiology and Parasitology, Split University Hospital and School of Medicine, Split, Croatia), Veronica Hola (Institute for Microbiology Masaryk, University Pekarska, Brno, Czech Republic), Piero Marone (Microbiology and Virology Department, Fondazione IRCCS Policlinico San Matteo, Pavia, Italy), Maria Teresa Mascellino (Dip. Sanita' Pubblica E Malattie Infettive, Sapienza University, Rome, Italy), Harald Seifert (Institute for Medical Microbiology, Immunology and Hygiene, University of Cologne, Cologne, Germany), Sonja Swidsinski (Department of Microbiology, Vivantes Hospital, Berlin, Germany) for kindly providing us with *Acinetobacter baumannii* clinical isolates. Italian authors are also indebted to Antonino Salvia, Director of Medical Services of the Fondazione Santa Lucia in Rome, for the useful information and advice on the clinical issues of this research.

Transparency Declaration
The authors declare no conflicts of interest.

Financial Support
This study was partially funded by the ESCMID Study Group for Biofilms with the Research Grant 5833 assigned to Gianfranco Donelli and Luisa Peixe.

References

Adams-Haduch JM, Onuoha EO, Bogdanovich T, Tian GB, Marschall J, Urban CM, Spellberg BJ, Rhee D, Halstead DC et al (2011) Molecular epidemiology of carbapenem-non susceptible *Acinetobacter baumannii* in the United States. J Clin Microbiol 49:3849–3854

Antunes LCS, Visca P, Towner KJ (2014) *Acinetobacter baumannii:* evolution of a global pathogen. Pathog Dis 71:292–301

Bartual SG, Seifert H, Hippler C, Luzon MA, Wisplinghoff H, Rodríguez-Valera F (2005) Development of a multi locus sequence typing scheme for characterization of clinical isolates of *Acinetobacter baumannii*. J Clin Microbiol 43:4382–4390

CLSI. Performance standards for antimicrobial susceptibility testing; twentieth informational supplement (2011) CLSI document M100-S21. Clinical and Laboratory Standards Institute, Wayne

Dettori M, Piana A, Deriu MG, Lo Curto P, Cossu A, Musumeci R et al (2014) Outbreak of multidrug-resistant *Acinetobacter baumannii* in an intensive care unit. New Microbiol 37:185–191

Djeribi R, Bouchloukh W, Jouenne T, Menaa B (2012) Characterization of bacterial biofilms formed on urinary catheters. Am J Infect Control 40:854–859

Donelli G, Vuotto C, Cardines R, Mastrantonio P (2012) Biofilm-growing intestinal anaerobic bacteria. FEMS Immunol Med Microbiol 65:318–325

Eijkelkamp BA, Stroeher UH, Hassan KA, Papadimitrious MS, Paulsen IT, Brown MH (2011) Adherence and motility characteristics of clinical *Acinetobacter baumannii* isolates. FEMS Microbiol Lett 323:44–51

Espiral P, Martì S, Vila J (2014) Effect of biofilm formation on the survival of *Acinetobacter baumannii* on dry surfaces. J Hosp Infect 80:56–60

Farshadzadeh Z, Hashemi FB, Rahimi S, Pourakbari B, Esmaeili D, Haghighi MA, Majidpour A, Shojaa S, Rahmani M et al (2015) Wide distribution of carbapenem resistant *Acinetobacter baumannii* in burns patients in Iran. Front Microbiol 6:1146

Fu Y, Zhou J, Zhou H, Yang Q, Wei Z, Yu Y, Li L (2010) Wide dissemination of OXA-23-producing carbapenem-resistant *Acinetobacter baumannii* clonal complex 22 in multiple cities of China. J Antimicrob Chemother 65:644–650

Giannouli M, Antunes LC, Marchetti V, Triassi M, Visca P, Zarrilli R (2013) Virulence-related traits of epidemic *Acinetobacter baumannii* strains belonging to the international clonal lineages I-III and to the emerging genotypes ST25 and ST78. BMC Infect Dis 13:282

Gniadek TJ, Carroll KC, Simner PJ (2016) Carbapenem-resistant non-glucose-fermenting gram-negative bacilli: the missing piece to the puzzle. J Clin Microbiol 54:1700–1710

Grosso F, Quinteira S, Peixe L (2011) Understanding the dynamics of imipenem-resistant *Acinetobacter baumannii* lineages within Portugal. Clin Microbiol Infect 17:1275–1279

Grosso F, Quinteira S, Poirel L, Novais A, Peixe L (2012) Role of common $bla_{OXA-24/OXA-40}$-carrying platforms and plasmids in the spread of OXA-24/OXA-40 among *Acinetobacter* species clinical isolates. Antimicrob Agents Chemother 56:3969–3972

Gundi VA, Dijkshoorn L, Burignat S, Raoult D, La Scola B (2009) Validation of partial *rpoB* gene sequence analysis for the identification of clinically important and emerging *Acinetobacter* species. Microbiology 155:2333–2341

Jones RN, Flonta M, Gurler N, Cepparulo M, Mendes RE, Castanheira M (2014) Resistance surveillance program report for selected European nations (2011). Diagn Microbiol Infect Dis 78:429–436

Kohlenberg A, Brümmer S, Higgins PG, Sohr D, Piening BC, de Grahl C, Halle E, Rüden H, Seifert H (2009) Outbreak of carbapenem-resistant *Acinetobacter baumannii* carrying the carbapenemase OXA-23 in a German university medical centre. J Med Microbio 58:1499–1507

Liu LL, Ji SJ, Ruan Z, Fu Y, YQ F, Wang YF, YS Y (2015) Dissemination of blaOXA-23 in *Acinetobacter spp.* in China: main roles of conjugative plasmid pAZJ221 and transposon Tn2009. Antimicrob Agents Chemother 59:1998–2005

Longo F, Vuotto C, Donelli G (2014) Biofilm formation in *Acinetobacter baumannii*. New Microbiol 37:119–127

Luo TL, Rickard AH, Srinivasan U, Kaye KS, Foxman B (2015) Association of bla_{OXA-23} and bap with the persistence of *Acinetobacter baumannii* within a major healthcare system. Front Microbiol 6:182

McConnell MJ, Actis L, Pachón J (2013) *Acinetobacter baumannii*: human infections, factors contributing to pathogenesis and animal models. FEMS Microbiol Rev 37:130–155

Mosqueda N, Gato E, Roca I, López M, de Alegría CR, Fernández Cuenca F, Martínez-Martínez L, Pachón J, Cisneros JM et al (2014) Characterization of plasmids carrying the blaOXA-24/40 carbapenemase gene and the genes encoding the AbkA/AbkB proteins of a toxin/antitoxin system. J Antimicrob Chemother 69:2629–2633

Peleg AY, Seifert H, Paterson DL (2008) Acinetobacter baumannii: emergence of a successful pathogen. Clin Microbiol Rev 21:538–582

Poirel L, Nordmann P (2006) Carbapenem resistance in Acinetobacter baumannii: mechanisms and epidemiology. Clin Microbiol Infect 12:826–836

Principe L, Piazza A, Giani T, Bracco S, Caltagirone MS, Arena F, Nucleo E, Tammaro F, Rossolini GM, et al; AMCLI-CRAb Survey Participants (2014) Epidemic diffusion of OXA-23-producing *Acinetobacter baumannii* isolates in Italy: results of the first cross-sectional countrywide survey. J Clin Microbiol 52:3004–3010

Qi L, Li H, Zhang C, Liang B, Li J, Wang L, Du X, Liu X, Qiu S et al (2016) Relationship between antibiotic resistance, biofilm formation, and biofilm-specific resistance in *Acinetobacter baumannii*. Front Microbiol 7:483

Rao RS, Karthika RU, Singh SP, Shashikala P, Kanungo R, Jayachandran S, Prashanth K (2008) Correlation between biofilm production and multiple drug resistance in imipenem resistant clinical isolates of *Acinetobacter baumannii*. Indian J Med Microbiol 26:333–337

Rodríguez-Baño J, Martí S, Soto S, Fernández-Cuenca F, Cisneros JM, Pachón J, Pascual A, Martínez-Martínez L, McQueary C, et al; Spanish Group for the Study of Nosocomial Infections (GEIH) (2008) Biofilm formation in *Acinetobacter baumannii*: associated features and clinical implications. Clin Microbiol Infect 14:276–278

Rolain JM, Loucif L, Al-Maslamani M, Elmagboul E, Al-Ansari N, Taj-Aldeen S, Shaukat A, Ahmedullah H, Hamed M (2016) Emergence of multidrug-resistant *Acinetobacter baumannii* producing OXA-23 Carbapenemase in Qatar. New Microbes New Infect 11:47–51

Roux D, Danilchanka O, Guillard T, Cattoir V, Aschard H, Fu Y, Angoulvant F, Messika J, Ricard JD et al (2015) Fitness cost of antibiotic susceptibility during bacterial infection. Sci Transl Med 7:297ra114

Sahl JW, Del Franco M, Pournaras S, Colman RE, Karah N, Dijkshoorn L, Zarrilli R (2015) Phylogenetic and genomic diversity in isolates from the globally distributed *Acinetobacter baumannii* ST25 lineage. Sci Rep 5:15188

Seifert H, Dolzani L, Bressan R, van der Reijden T, van Strijen B, Stefanik D, Heersma H, Dijkshoorn L (2005) Standardization and interlaboratory reproducibility assessment of pulsed-field gel electrophoresis-generated fingerprints of *Acinetobacter baumannii*. J Clin Microbiol 43:4328–4335

Turton JF, Gabriel SN, Valderrey C, Kaufmann ME, Pitt TL (2007) Use of sequence-based typing and multiplex PCR to identify clonal lineages of outbreak strains of *Acinetobacter baumannii*. Clin Microbiol Infect 13:807–815

Viana GF, Zago MC, Moreira RR, Zarpellon MN, Menegucci TC, Cardoso CL, Tognim MC (2016) ISAba1/blaOXA-23: a serious obstacle to controlling the spread and treatment of *Acinetobacter baumannii* strains. Am J Infect Control 44:593–595

Vuotto C, Donelli G (2014) Field emission scanning electron microscopy of biofilm-growing bacteria involved in nosocomial infections. Methods Mol Biol 1147:73–84

Woodford N, Ellington MJ, Coelho JM, Turton JF, Ward ME, Brown S, Amyes SG, Livermore DM (2006) Multiplex PCR for genes encoding prevalent OXA carbapenemases in *Acinetobacter spp.* Int J Antimicrob Agents 27:351–353

Wu W, He Y, Lu J, Lu Y, Wu J, Liu Y (2015) Transition of blaOXA-58-like to blaOXA-23-like in *Acinetobacter baumannii* clinical isolates in Southern China: an 8-year study. PLoS One 10:e0137174

Zarrilli R (2016) *Acinetobacter baumannii* virulence determinants involved in biofilm growth and adherence to host epithelial cells. Virulence 7:367–368

Zarrilli R, Pournaras S, Giannouli M, Tsakris A (2013) Global evolution of multidrug-resistant Acinetobacter Baumannii clonal lineages. Int J Antimicrob Agents 41:11–19

Adv Exp Med Biol - Advances in Microbiology, Infectious Diseases and Public Health (2018) 9: 85–94
DOI 10.1007/5584_2017_97
© Springer International Publishing AG 2017
Published online: 13 September 2017

Pragmatic Combination of Available Diagnostic Tools for Optimal Detection of Intestinal Microsporidia

Stuti Kaushik, Rumpa Saha, Shukla Das, VG Ramachandran, and Ashish Goel

Abstract

Diarrhea is a debilitating condition in HIV infected individuals and with the finding that almost 1/4 cases of diarrhea in HIV are due to microsporidia, there is a dire need to institute measures for its detection on a regular basis. Keeping this in mind the study aims to determine the burden of intestinal microsporidiosis in HIV seropositive patients presenting with and without diarrhea and to compare the ability of microscopy and PCR in its detection.

The study group consisted of 120 patients divided into four groups HIV seropositive with/without diarrhea, and HIV seronegative with/without diarrhea. Performance of four staining techniques including Modified Trichrome, Calcofluor White, Gram Chromotrope and Quick hot Gram Chromotrope stains were evaluated against PCR in diagnosing enteric microsporidiosis from stool samples.

Overall prevalence of intestinal microsporidiosis was 10.83%. The same for HIV seropositive patients with diarrhea was 23.33%, HIV seropositive patients without diarrhea and in immune-competent hosts with diarrhea was 10% each. *Enterocytozoon bieneusi* was found to predominate. Calcofluor white stain detected maximum microsporidia in stool samples (76.92%), followed by Modified Trichrome stain (61.5%), PCR (46.15%) and Gram Chromotrope and Quick hot Gram Chromotrope stains (38.4% each). PCR exhibited the best performance with a sensitivity and specificity of 100%. Our data suggests screening of stool samples with either Modified Trichrome or Calcofluor white stain followed by PCR confirmation thus leading to maximum detection along with speciation for complete cure.

Keywords

AIDS · Albendazole · Diarrhea · *Enterocytozoon* · Microsporidia

S. Kaushik, R. Saha (✉), S. Das, and V. Ramachandran
Department of Microbiology, University College of Medical Sciences & Guru Teg Bahadur Hospital, Dilshad Garden, Delhi 110095, India
e-mail: stuti27@gmail.com; rumpachatterjee@yahoo.co.in; shukladas_123@yahoo.com; rama_88@yahoo.com

A. Goel
Department of Medicine, University College of Medical Sciences & Guru Teg Bahadur Hospital, Dilshad Garden, Delhi 110095, India
e-mail: ashgoe@gmail.com

1   Introduction

Ever since the first documentation of human microsporidiosis in 1959 in Japan, these agents have been linked with opportunistic infections in patients' immunosuppressed due to HIV (Matsubayashi et al. 1959; Didier and Weiss 2006; Lee et al. 2010). Human infection is known to be caused by 17 different species of microsporidia belonging to eight genera (Candramathi et al. 2012). Besides HIV, immunosuppression due to other causes also predisposes to infection, with recent studies highlighting the occurrence in organ transplant recipients (Ghoshal et al. 2015a). Immunocompetent subjects have not been spared either (Didier 2005).

Intestinal microsporidiosis contributes to significant morbidity and mortality in HIV infected patients. The most common manifestation in them is chronic diarrhea and a wasting syndrome (Weber et al. 1994; Anane and Attouchi 2010). *Enterocytozoon bieneusi* and *Encephalitozoon intestinalis* are notorious in enteric microsporidiosis. Of the two aforementioned species, *E.bieneusi* is more common in chronic diarrhea while infections due to *E.intestinalis* tend to disseminate. Gastrointestinal infections in immunocompetent hosts are self- limiting (Barratt et al. 2010).

Vertebrate hosts are now identified for all four major microsporidia species infecting humans, (*E.bieneusi* and three *Encephalitozoon* species) implying a zoonotic nature of these agents (Mathis et al. 2005). This finding warrants evaluation of seroprevalence of the 4 most commonly involved species causing human diseases in a tropical country like India to (i) find how common the exposure is in healthy populations and (ii) to elucidate whether this organism is present in the human host in a state of dormancy and undergoes recrudescence in conditions of major or minor immunosuppression.

The diagnosis of this agent has traditionally depended on microscopy. Identification of the environmentally resistant spore that bears an internal polar filament forms the basis for diagnosis by microscopy. The major drawback of microscopy is the small size of the spore $(1–1.5 \times 2 – 4\ \mu)$ which is hard to be noticed by untrained eye and depends on the expertise of the microscopist.

Non-availability of specific antibodies against microsporidia has limited the development of an assay for antigen detection. Antibodies against *Encephalitozoon spp.* have been demonstrated in the serum of patients but this evidence in itself did not conclusively discriminate true infections and cross reactivity (Fedorko and Hijazi 1996). Cell culture is inappropriate for diagnosing enteric microsporidiosis (Visvesvara et al. 1995). Limited homology of single sub-unit (SSU) rRNA genes of microsporidia with eukaryotic organisms has allowed the utility of these genes for molecular techniques for diagnosis and speciation of intestinal mirosporidia (Zhu et al. 1993).

The prevalence of intestinal microsporidiosis ranges between 7 and 50% in the west and *E. bieneusi* is more prevalent as compared to *E. intestinalis*. In India microsporidia is a neglected agent of diarrhea, especially in the HIV infected patients and this is due to the paucity of proper diagnostic techniques. Limited studies that have been conducted show the prevalence to range between 2 and 26.7% and studies on species identity and predominance are scarce (Gupta et al. 2008; Tuli et al. 2008; Kulkarni et al. 2011; Saigal et al. 2013a; Ghoshal et al. 2015a).

In this backdrop the present study was conducted to determine the burden of intestinal microsporidiosis in patients presenting with and without diarrhea in HIV seropositive individuals and to compare the ability of traditional staining methods and PCR in detection of intestinal microsporidiosis.

2   Material and Methods

This cross-sectional observational study was conducted in our tertiary care hospital in Delhi between November 2014 and April 2016 after institutional ethics clearance and written consent.

A total of 120 patients, all >12 years were included in the study and were further divided into four groups: 30 each HIV seropositive treatment naive patients with/ without diarrhea and 30 each HIV seronegative patients with/without diarrhea. A detailed history regarding the demography and the symptoms pertaining to diarrhea was taken and recorded.

Subjects with history of intake of antiparasitic drugs, antibiotics or antimotility drugs in preceding 2 weeks were excluded.

Fresh stool samples were collected and subjected to routine processing after concentration using formalin ether sedimentation method. Smears were prepared and fixed with absolute methanol for 2–3 min for modified trichrome stain and Calcofluor White but heat fixed for gram chromotrope, quick hot gram chromotrope and modified Ziehl Neelsen stains (Garcia 2002).

DNA was extracted from stool samples stored at −20 °C using Qiagen (Germany) 'QIAamp DNA Stool Mini Kit' according to manufacturer's protocol and stored at −20 °C. PCR was done using primers targeting a conserved region of SSU rRNA gene of four common intestinal microsporidia: *E.bieneusi, E. intestinalis, E.cuniculi* and *E.hellem* (Forward C1:5′CACCAGGTTGATTCTGCC3′ and Reverse C2: 5′GTGACGGGCGGTGTGTAC3′). The amplification protocol included an initial denaturation at 94 °C for 5 min (min) followed by 30 cycles of denaturation at 94 °C for 1 min, annealing at 56 °C for 1 min, elongation at 72 °C for 1 min and 5 min of extension at 72 °C after 30 cycles (Raynaud et al. 1998). The samples showing a specific band of 1200 bp were subjected to PCR for speciation. Species determination for *E.bieneusi* (V1 Forward: 5′CACCAGGTTGATTC-TGCCTGA3′ and EB450 Reverse: 5′ACTCAGGTGTTATACTCACGTC3′) and *E. intestinalis* (V1 Forward: 5′CACCAGGTTGA-TTCTGCCTGA3′ and SI500 Reverse: 5′CTCGCTCCTTTACACTCGAA3′) were done using species specific primers (Coyle et al. 1996). Cycling conditions included an initial

Table 1 Interpretation of kappa value

Kappa value	Agreement
< 0	Less than chance agreement
0.01–0.20	Slight agreement
0.21–0.40	Fair agreement
0.41–0.60	Moderate agreement
0.61–0.80	Substantial agreement
0.81–0.99	Almost perfect agreement

denaturation of DNA at 94 °C for 5 min followed by 35 cycles of denaturation at 94 °C for 1 min, annealing at 50.2 °C for 1 min (*E.bieneusi*) / 58 °C for 1 min (*E intestinalis*), elongation at 72 °C for 1 min and 5 min of extension at 72 °C after 35 cycles.

Chi Square test and unpaired Student T test were used to compare proportion between various groups. McNemar's test was used for intra group comparison. Sensitivity, specificity and predictive values were calculated using standard formulae. Cohen's Kappa test was used for assessment of interrater reliability. Interpretation of Kappa value is shown in Table 1 (Viera and Garrett 2005).

3   Results

Among the 120 subjects included in the study, 10.8% (13/120) stool samples were found to be positive for microsporidia by any of the detection methods used in the study. Microsporidia spores were detected in a significant number of cases 14% (13/90) and in 0% of the controls. ($P = 0.036$).

The mean age of individuals with intestinal microsporidiosis was 30 years. The most common age group presenting with gastrointestinal microsporidiosis was 29–36 years but no significant association of intestinal microsporidiosis with age was observed. ($P = 0.24$) Majority of the patients with intestinal microsporidiosis were males (M: F = 8:5, $P = 0.51$).

None of the demographic parameters assessed (including access to filtered drinking water) was significantly associated with intestinal microsporidiosis.

Although 80% of the patients with intestinal microsporidiosis had acute diarrhea, nausea and vomiting were found to be significantly associated both in immunosuppressed and immunocompetent individuals. ($P = 0.0006$).

HIV seropositive patients with intestinal microsporidiosis were equally divided in stage I and III of WHO clinical stage of AIDS.

Although blood and mucus were absent in the stool of patients with intestinal microsporidiosis, significant anemia ($P = 0.005$) was noted in these patients.

Table 2 depicts the number of stool samples that were positive for spores of microsporidia using different diagnostic modalities employed in the study. Figure 1a, b, c, and d depicts the spores of microsporidia on modified trichrome, Calcofluor White, gram chromotrope and quick hot gram chromotrope stain.

Six stool specimens were positive for microsporidia by PCR utilizing pan-specific primers (Fig. 2a). Four of these six were *E. bieneusi* and the remaining two were not amplified with either *E.bieneusi* or *E.intestinalis* specific primers, remaining unspeciated. (Fig. 2b). *E. bieneusi* was found in 3 HIV seropositive and 1 HIV seronegative patient respectively. Both the unspeciated microsporidia were

detected in HIV seronegative patients with diarrhea.

Six (5%) samples were positive by 2 or more techniques and these were considered true positives and 107 (90%) samples which were negative by all five of the above mentioned techniques were considered as true negatives as shown in Table 3.(Saigal et al. 2013b).

Sensitivity and specificity of the diagnostic methods were calculated using true positives and true negatives (Table 4). The inter-rater reliability of various methods is specified in Table 5.

PCR was found to be 100% sensitive and specific and in almost perfect agreement with gram chromotrope, quick hot gram chromotrope and modified trichrome staining.

4 Discussion

The present study showed a 10.8% (13/120) prevalence of intestinal microsporidiosis in Delhi. This increased to 16.7% (10/60) in HIV seropositive patients, comparable to a study from Chandigarh (15.9%), (Saigal et al. 2013a). A lower prevalence of 2% in HIV seropositive individuals has been reported from Gujarat (Gupta et al. 2008).

Microsporidia detection in HIV seropositive individuals with diarrhea was 23.3% (7/30)

Table 2 Identification of microsporidia by different diagnostic modalities

Method of detection	HIV reactive			HIV non- reactive			Total Microsporidia detected by each method
	Diarrhea present (n = 30)	Diarrhea absent (n = 30)	Total (n = 60)	Diarrhea present (n = 30)	Diarrhea absent (n = 30)	Total (n = 60)	
Staining							
Modified trichrome	3	2	5	3	0	3	8(61.5%)
Calcofluor white	5	3	8	2	0	2	10(76.92%)
Gram chromotrope	1	2	3	2	0	2	5(38.4%)
Quick hot gram chromotrope	1	2	3	2	0	2	5(38.4%)
PCR (Pan-specific)	1	2	3	3	0	3	6(46.15%)

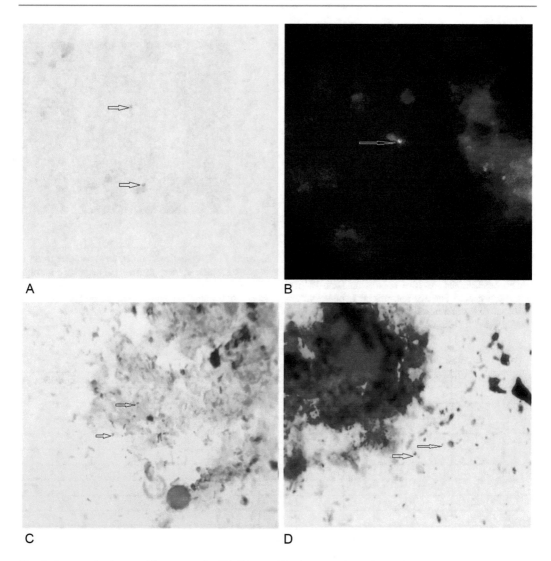

Fig. 1 Spores of microsporidia seen under X1000 magnification on (**a**) Modified trichrome stain, (**b**) Calcofluor White stain under fluorescent microscope, (**c**) Gram chromotrope stain, (**d**) Quick hot gram chromotrope stain

which is higher than reports from Mumbai (17.8%). (Dalvi et al. 2006;). Chandigarh (30.4%) and Varanasi (26.7%) documents even higher rates (Tuli et al. 2008; Saigal et al. 2013a).

The prevalence of intestinal microsporidiosis in HIV seropositive patients without diarrhea was 10% (3/30), akin to report from Chandigarh (7.6%), (Saigal et al. 2013a).

Microsporidiosis is reported in immune-competent hosts in recent times. A prevalence of 10% (3/30) was noted in our immune-competent patients analogous to reports from

India (10.9%) and Turkey (9.8%), (Saigal et al. 2013a; Türk et al. 2012). A lower rate of 1.2% and 5% is reported from North India and Pakistan, unlike disproportionately higher rates of 67.5% and 38% documented from Cameroon and Czech Republic in healthy individuals (Ghoshal et al. 2016; Yakoob et al. 2012; Nkinin et al. 2007; Sak et al. 2011). These studies indicate that immune-competent hosts are in fact at increased risk of infection with intestinal microsporidia.

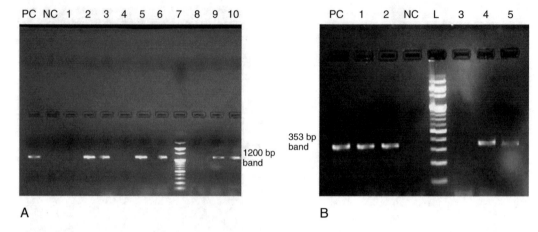

Fig. 2 (**a**) An agarose gel showing PCR products of human fecal samples using Pan specific microsporidia primer; positive and negative control on Lane PC and NC; microsporidia positive samples on Lane 2, 3, 5, 6, 9, 10; microsporidia negative samples on Lane 1, 4, 8, 100 bp DNA ladder on Lane 7. (**b**) An agarose gel showing PCR products of human fecal samples using *E. bieneusi* specific primer; positive and negative control on Lane PC and NC; *E.bieneusi* positive samples on Lane 1, 2, 4, 5; *E.bieneusi* negative samples on Lane 3, 100 bp DNA ladder on Lane L

Table 3 Estimation of true positives and true negatives

Sample Number	MTS	CFW	GC	QHGC	PCR	Number of methods by which microsporidia detected
1	+	−	−	−	−	1
2	+	−	−	−	−	1
3	−	+	−	−	−	1
4	+	+	+	+	+	5
5	+	+	+	+	+	5
6	−	+	−	−	−	1
7	−	+	−	−	−	1
8	−	+	−	−	−	1
9	+	+	−	−	+	3
10	−	+	−	−	−	1
11	+	+	+	+	+	5
12	+	+	+	+	+	5
13	+	−	+	+	+	4
14–120	−	−	−	−	−	Not detected by any method
120	**8**	**10**	**5**	**5**	**6**	

MTS Modified Trichrome stain, *CFW* calcofluor white, *GC* Gram chromotrope, *QHGC* Quick hot gram chromotrope

Table 4 Comparison of staining and PCR in detection of microsporidia

Method of detection of Microsporidia	Sensitivity (%)	Specificity (%)	Positive predictive value (%)	Negative predictive value (%)
PCR	100	100	100	100
Modified trichrome stain	100	98.24	75	100
Calcofluor white stain	83.33	95.61	50	99.09
Gram chromotrope stain	83.33	100	100	99.13
Quick hot gram chromotrope stain	83.33	100	100	99.13

Table 5 Inter rater reliability of diagnostic modalities for detection of Microsporidia in stool

	MTS +	MTS -	Inter-rater reliability (kappa value)	Interpretation
PCR +	6	0	0.848	**Almost perfect agreement**
PCR -	2	112		
	CFW +	**CFW -**		
PCR +	5	1	0.600	Substantial agreement
PCR -	5	109		
	GC+	**GC-**		
PCR +	5	1	0.905	**Almost perfect agreement**
PCR -	0	114		
	QHGC	**QHGC**		
PCR +	5	1	0.905	**Almost perfect agreement**
PCR -	0	114		
	CFW +	**CFW -**		
MTS +	5	3	0.520	Moderate agreement raynaud
MTS -	5	107		
	GC +	**GC -**		
MTS +	5	3	0.757	Substantial agreement
MTS -	0	112		
	QHGC+	**QHGC-**		
MTS +	5	3	0.757	Substantial agreement
MTS -	0	112		
	GC +	**GC -**		
CFW+	4	6	0.506	Moderate agreement
CFW -	1	109		
	QHGC +	**QHGC -**		
CFW+	4	6	0.506	Moderate agreement
CFW-	1	109		
	QHGC +	**QHGC -**		
GC +	5	0	1.00	Almost perfect agreement
GC -	0	115		

MTS Modified Trichrome stain, *CFW* calcofluor white, *GC* Gram chromotrope, *QHGC* Quick hot gram chromotrope

The varying rate of prevalence across different countries is perhaps a reflection of different techniques used for detection. Whether geographical and ethnic differences play any role or not needs further investigation.

The mean age of immunocompromised patients with intestinal microsporidiosis was ~30 yr which is comparable to a report from India (Ghoshal et al. 2015a).

The presentation of acute diarrhea in 80% of our intestinal microsporidiosis patients contrasts with finding by Ujjala et al. who found chronic diarrhea in 92.3% of their enteric

microsporidiosis patients (Ghoshal et al. 2015a). Our findings can be explained by early presentation and timely collection of sample from HIV seropositive patients attending to the ART (anti retroviral treatment) clinic.

The stool was watery in 70% (7/10) cases of intestinal microsporidiosis with diarrhea. This corroborates well with previously published reports that the consistency of stool becomes liquid with increased shedding of oocysts (Dalvi et al. 2006).

Inflammation as indicated by the presence of pus cells was not prominent with microsporidiosis which is consistent with South African study probably suggesting that inflammatory response and intestinal cell injury are minimal in gastrointestinal microsporidiosis (Samie et al. 2007).

No association was seen between intestinal microsporidiosis and clinical stage of AIDS at presentation suggesting that enteric microsporidiosis can be acquired at any stage of clinical AIDS.

Contrary to almost all previous studies in India and Iran who have reported significant association with low CD4 counts (at ≤ 200/cumm), the present study found no association between the two (Tuli et al. 2008; Dalvi et al. 2006; Agholi et al. 2013). Reports from the developed world also identified microsporidia in HIV infected individuals with comparatively well preserved CD4 levels (Cimerman et al. 1999; Bryan and Schwartz 1999). The understanding of immunological response to microsporidial infections in humans is based on knockout gene models in mice, infected with *E. cuniculi*. It has been shown that the cardinal role in clearing this organism is accomplished by CD8 T-cells, which is enhanced by IFN-γ. Mice devoid of CD4 are able to survive high dose infection while CD8 gene knockout mice cannot survive infection. The zealous use of CD4 cell counts to monitor the progression of AIDS has made it difficult to conclude the relation between microsporidial infection and CD4 and CD8 counts (Khan et al. 2001).

Majority (66.7%) and all HIV seropositive cases of intestinal microsporidiosis in the present study were found to be due to *E.bieneusi*, enlisting it to be the predominant species in Delhi and reiterating its predominance in HIV seropositive individuals. *E.bieneusi* was found in 33.3% of the HIV seronegative patients with diarrhea. *E.intestinalis* was not detected in any sample. Our findings are similar to studies from Lucknow (100%) and Pune (60%), but unlike that from Chandigarh where *E.intestinalis* has been found to predominate, both in HIV seropositive and HIV seronegative individuals (Ghoshal et al. 2015a; Kulkarni et al. 2011; Saigal et al. 2013a). *E.bieneusi* also predominates in Nigeria

and Vietnam (Akinbo et al. 2010; Espern et al. 2007). Thus arises the importance of species determination in this genus, since *E.bieneusi* exhibits intrinsic resistance to albendazole, which is conventionally used in the treatment of intestinal microsporidiosis and preventing its dissemination in advanced stages of the disease.

In the present study, 33.3% cases of microsporidiosis (all immune-competent) remained unspeciated due to limitations of the methodology employed. They could either be due to *E.cuniculi* or *E.hellem*. No studies in India have yet reported either of them as enteric pathogens in immunosuppressed or in immunocompetent hosts. *E.cuniculi* and *E.hellem* have been reported in humans in the developed world in HIV seropositive patients (Weber et al. 1994). Enteric localization of *E.cuniculi* as a part of disseminated disease has also been reported in two immunocompromised cases (Fournier et al. 2000).

In our effort to compare the 5 diagnostic modalities, it was found that PCR and modified trichrome stain were 100% sensitive. However, specificity of modified trichrome stain (98.2%) was unlike PCR (100%), (Table 4). Sensitivity and specificity of modified trichrome stain has been reported to vary from 64 to 94% and 100% respectively in two recent studies from north India (Saigal et al. 2013b; Ghoshal et al. 2015b). Great technical expertise is required for identifying the minute (1–3 µ) spores of this organism. Other drawbacks of this stain are a shorter shelf life of prepared stain and frequent eye strain during screening.

Specificity and positive predictive value (PPV) was 100% with gram chromotrope and quick hot gram chromotrope stains. However the violet colored oval spores can be easily missed against a reddish background, thereby lowering the sensitivity of these methods. All the above staining methods are time consuming, labor intensive and require expertise.

Although sensitivity (83.3%) and specificity (95.6%) of Calcofluor White was low, this stain had the advantage of lesser staining time, ease of staining, higher shelf life coupled with easier identification. Although the maximum numbers of spores of microsporidia were identified by this method (76.9%), (Table 2), the PPV still remains low (50%). The stain binds to other chitin bearing fungal structures which may account for the lower PPV. Rapid loss of fluorescence on exposure to light and increased number of false positives identified was major shortcomings. Technical expertise with a fluorescent microscope needs to be kept in mind.

PCR, the only method with 100% sensitivity, specificity and PPV and coupled with speciation (by using species specific primers), attains therapeutic and epidemiological importance.

PCR had almost perfect agreement with modified trichrome stain, gram chromotrope and quick hot gram chromotrope stains (Table 5), but due to non-contrasting background of the latter two stains, they are not preferable for microscopy (Sensitivity 83.3%). Modified trichrome stain at the same time will differentiate the spore better due to the contrasting background (Sensitivity 100%).

Our data suggests screening of stool samples with either modified trichrome or Calcofluor White followed by PCR confirmation and speciation leading to maximum detection.

Keeping in mind the well known fact that Microsporidia are shed intermittently in feces, a single sample and low case load in each group with its inevitable effect on prevalence and statistical analysis are drawbacks of the present study which were unavoidable due to poor compliance of patients and cost constraints. Additionally inclusion of only adults prevented assessment of this clinical entity in individuals less than 12 years of age. Also amongst the patients included in the study, only four were above the age of 60 years. Hence an assessment of increased vulnerability to enteric microsporidiosis at extremes of age will require studies targeting these age groups. Follow up of the cases of gastrointestinal microsporidiosis upon the resolution of disease after treatment was not possible, due to poor patient compliance, precluding a comment correlating treatment and resolution of the disease.

References

Agholi M, Hatam GR, Motazedian MH (2013) HIV/AIDS-associated opportunistic protozoal diarrhea. AIDS Res Hum Retrovir 29(1):35–41

Akinbo FO, Okaka CE, Omoregie R (2010) Prevalence of intestinal parasitic infections among HIV patients in Benin City, Nigeria. Libyan J Med 5:5506

Anane S, Attouchi H (2010) Microsporidiosis: epidemiology, clinical data and therapy. Gastroenterol Clin Biol 34:450–464

Barratt JL, Harkness J, Marriott D, Ellis JT, Stark D (2010) Importance of nonenteric protozoan infections in immunocompromised people. Clin Microbiol Rev 23(4):795–836

Bryan RT, Schwartz DA (1999) Epidemiology of microsporidiosis. In: Witner M, Weiss LM (eds) The microsporidia and microsporidiosis. ASM Press, Washington, DC, pp 502–516

Candramathi S, Suresh K, Anita ZA, Kuppusamy UR (2012) Infections of blastocystis hominis and microsporidia in cancer patients: are they opportunistic? Trans R Soc Trop Med Hyg 106:267–269

Cimerman S, Cimerman B, Lewi DS (1999) Prevalence of intestinal parasitic infections in patients with acquired immunodeficiency syndrome in Brazil. Int J Infect Dis 3(4):203–206

Coyle CM, Wittner M, Kotler DP, Noyer C, Orenstein JM, Tanowitz HB, Weiss LM (1996) Prevalence of microsporidiosis due to *Enterocytozoon bieneusi* and *Encephalitozoon (Septata) intestinalis* among patients with AIDS-related diarrhea: determination by polymerase chain reaction to the microsporidial small-subunit rRNA gene. Clin Infect Dis 23(5):1002–1006

Dalvi S, Mehta P, Koticha A, Gita N (2006) Microsporidia as an emerging cause of parasitic diarrhea in HIV seropositive individuals in Mumbai. Bombay Hosp J 48:592–597

Didier ES (2005) Microsporidiosis: an emerging and opportunistic infection in humans and animals. Acta Trop 94:61–76

Didier ES, Weiss LM (2006) Microsporidiosis: current status. Curr Opin Infect Dis 19(5):485–492

Espern A, Morio F, Miegeville M, Illa H, Abdoulaye M, Meyssonnier V, Adehossi E, Lejeune A, Cam PD, Besse B, Gay-Andrieu F (2007) Molecular study of microsporidiosis due to *Enterocytozoon bieneusi* and *Encephalitozoon intestinalis* among human immunodeficiency virus-infected patients from two geographical areas: Niamey, Niger, and Hanoi, Vietnam. J Clin Microbiol 45(9):2999–3002

Fedorko DP, Hijazi YM (1996) Application of molecular techniques to the diagnosis of microsporidial infection. Emerg Infect Dis 2:183–191

Fournier S, Liguory O, Sarfati C, David-Ouaknine F, Derouin F, Decazes J, Molina J (2000) Disseminated infection due to *Encephalitozoon cuniculi* in a patient with AIDS: case report and review. HIV Med 1 (3):155–161

Garcia LS (2002) Laboratory identification of the microsporidia. J Clin Microbiol 40:1892–1901

Ghoshal U, Khanduja S, Pant P, Prasad KN, Dhole TN, Sharma RK, Ghoshal UC (2015a) Intestinal microsporidiosis in renal transplant recipients: prevalence, predictors of occurrence and genetic characterization. Indian J Med Microbiol 33:357–363

Ghoshal U, Khanduja S, Agarwal V, Dhole TN, Ghoshal UC (2015b) Comparative evaluation of staining techniques and polymerase chain reaction for diagnosis of intestinal microsporidiosis in immunocompromised patients. Trop Parasitol 5:101–105

Ghoshal U, Dey A, Ranjan P, Khanduja S, Agarwal V, Ghoshal UC (2016) Identification of opportunistic enteric parasites among immunocompetent patients with diarrhea from Northern India and genetic characterization of *Cryptosporidium* and *Microsporidia*. Indian J Med Microbiol 34:60–66

Gupta M, Sinha M, Raizada N (2008) Opportunistic intestinal protozoan parasitic infection in HIV positive patients in Jamnagar, Gujrat. SAARC J Tuber Lung Dis HIV/AIDS 5(1):21–24

Khan IA, Moretto M, Weiss LM (2001) Immune response to *Encephalitozoon cuniculi* infection. Microbes Infect 3:401–405

Kulkarni S, Pastute S, Chandane M, Risbud A (2011) Performance of microscopy for detection of microsporidial spores from stool samples of HIV infected individuals with diarrhea. Indian J Med Res 134:982–984

Lee SC, Corradi N, Doan S, Dietrich FS, Keeling PJ, Heitman J (2010) Evolution of the sex-related locus and genomic features shared in microsporidia and fungi. PLoS One 5:e10539

Mathis A, Weber R, Deplazes P (2005) Zoonotic potential of the microsporidia. Clin Microbiol Rev 18 (3):423–445

Matsubayashi H, Koike T, Mikata T, Hagiwara S (1959) A case of Encephalitozoon like body infection in man. Arch Pathol 67:181–187

Nkinin SW, Asonganyi T, Didier ES, Kaneshiro ES (2007) Microsporidial infection is prevalent in healthy people in Cameroon. J Clin Microbiol 45(9):2841–2846

Raynaud L, Delbac F, Broussolle V, Rabodonirina M, Girault V, Wallon M, Cozon G, Vivares CP, Peyron F (1998) Identification of Encephalitozoon intestinalis in travelers with chronic diarrhea by specific PCR amplification. J Clin Microbiol 36(1):37–40

Saigal K, Sharma A, Sehgal R, Sharma P, Malla N, Khurana S (2013a) Intestinal microsporidiosis in India: a two year study. Parasitol Int 62(1):53–56

Saigal K, Khurana S, Sharma A, Sehgal R, Malla N (2013b) Comparison of staining techniques and multiplex nested PCR for diagnosis of intestinal microsporidiosis. Diag Microbiol Infect Dis 77:248–249

Sak B, Brady D, Pelikánová M, Květoňová D, Rost M, Kostka M, Tolarová V, Hůzová Z, Kváč M (2011) Unapparent microsporidial infection among immunocompetent humans in the Czech Republic. J Clin Microbiol 49(3):1064–1070

Samie A, Obi CL, Tzipori S, Weiss LM, Guerrant RL (2007) Microsporidiosis in South Africa: PCR detection in stool samples of HIV positive and HIV negative individuals and school children in Vhembe district, Limpopo Provinces. Trans R Soc Trop Med Hyg 101(6):547–554

Tuli L, Gulati AK, Sundar S, Mohapatra TM (2008) Correlation between CD4 counts of HIV patients and enteric protozoan in different seasons – an experience of a tertiary care hospital in Varanasi (India). BMC Gastroenterol 8:36

Türk S, Dogruman AF, Karaman U, Kustimur S (2012) Investigation of *Microsporidia* prevalence by different staining methods in cases of diarrhea. Microbiol Bull 46:85–92

Viera AJ, Garrett JM (2005) Understanding inter observer agreement: the kappa statistic. Fam Med 37 (5):360–363

Visvesvara GS, Leitch GJ, Pieniazek NJ, Da Silva AJ, Wallace S, Slemenda SB, Weber R, Schwartz DA, Gorelkin L, Wilcox CM (1995) Short term in vitro culture and molecular analysis of the microsporidian, *Enterocytozoon bieneusi*. J Euk Microbiol 42:506–510

Weber R, Bryan RT, Schwartz DA, Owen RL (1994) Human microsporidial infections. Clin Microbiol Rev 7:426–461

Yakoob J, Abbas Z, Beg MA, Jafri W, Naz S, Khalid A, Khan R (2012) Microsporidial infections due to *Encephalitozoon intestinalis*in in non-HIV-infected patients with chronic diarrhea. Epidemiol Infect 140:1773–1779

Zhu X, Wittner M, Tanowitz HB, Kotler D, Cali A, Weiss LM (1993) Small subunit rRNA sequence of *Enterocytozoon bieneusi* and its potential diagnostic role with use of the polymerase chain reaction. J Infect Dis 168:1570–1575

Adv Exp Med Biol - Advances in Microbiology, Infectious Diseases and Public Health (2018) 9: 95–100
DOI 10.1007/5584_2017_125
© Springer International Publishing AG 2017
Published online: 21 November 2017

The Fight Against Tuberculosis in the Mid-nineteenth Century: The Pivotal Contribution of Edoardo Maragliano (1849–1940)

Mariano Martini, Ilaria Barberis, Nicola Luigi Bragazzi, and Filippo Paluan

Abstract

The second half of the nineteenth century saw the development of new medical "specialties", which, like the idea of constitutional disease, had a profound influence on medical practice. Against this lively "backdrop", Edoardo Maragliano played a central role in medicine's "renaissance" in Italy. Having graduated in medicine in 1870 at the University of Naples, he worked as an assistant in the University Medical Clinic. After beginning his academic career as professor of pathology at the Faculty of Medicine in Genoa in 1877, he became full professor of internal medicine in 1881. While he studied all fields of internal medicine, his research focused mainly on tuberculosis.

His experiments in the medical clinic enabled Maragliano to announce the possibility of immunization against *Mycobacterium tuberculosis*. Although criticized for using an inactivated vaccine, Maragliano continued to advocate vaccination with any type of vaccine.

In the Senate of the Kingdom of Italy, Maragliano actively debated social, economic and sanitary questions, without neglecting his duties as a physician and professor. As an officer during the First World War, he organized military health services and taught medicine at the Military University of Padua.

In 1924, Maragliano created the first Italian specialty school in the study of tuberculosis, which provided physicians with specific training in the diagnosis, therapy and prevention of the disease. His scientific zeal and his vision of modern medicine prompted the introduction of new specializations, such as radiology and, especially, pneumology, which led to the creation of one of Europe's most renowned medical schools.

M. Martini
Department of Health Sciences, Section of History of Medicine and Ethics, University of Genoa, Genoa, Italy

I. Barberis and N.L. Bragazzi
Department of Health Sciences, University of Genoa, Genoa, Italy

F. Paluan (✉)
Department of Cardiac, Thoracic and Vascular Sciences, University of Padua, Padua, Italy
e-mail: filippo.pmz@gmail.com

Keywords

Edoardo Maragliano · History of
tuberculosis · Internal medicine · Public
health · Vaccination

Italian medicine played a central role in the
development of new sciences, through the
foundation of long-standing medical schools,
driven by such Masters of Medicine as Guido
Baccelli (1830–1916) in Rome, Augusto Murri
(1841–1932) in Bologna, Achille De Giovanni
(1824–1916) in Padua, Antonio Cardarelli
(1831–1927), Arnaldo Cantani (1837–1893),
Errico De Renzi (1839–1921) and Gaetano
Rummo (1853–1917) in Naples. It was in this
productive and stimulating period that Edoardo
Maragliano (1849–1940; Fig. 1) made his contribution to medical science (Borghi 2015;
Premuda 1975; Benedicenti 1940).

Maragliano was born of humble origins on 1st
June 1849, and began his medical studies in
Genoa, where he remained for 3 years. He subse-
quently completed his studies in Naples and
graduated in 1870, under the supervision of Professor Salvatore Tommasi (1813–1888) and

Fig. 1 Edoardo Maragliano in 1896 (Pizzini 1896)

Professor Arnaldo Cantani (1837–1893), after
which he became an assistant in the university's
medical clinic. On returning to Genoa, he began
work as an assistant in the university clinic,
which was directed by Professor Errico De
Renzi (1839–1921). In 1875, Ercole Galvagni
(1836–1909) and Edoardo Maragliano competed
for the post of full professor of internal medicine
and, on the basis of their qualifications and examination results, were declared joint winners; the
chair of medicine at the University of Cagliari
was assigned to Ercole Galvagni, because of his
senior age. While continuing to work in the
hospitals of his city, Maragliano began teaching
both students and physicians; from 1875 to 1880,
he taught neuropathology in the psychiatric hospital of Genoa (the so–called *Ospedale degli
Incurabili* [*Hospital for the Incurable*] or
Ospedale dei Cronici [*Hospital for Chronic
Patients*]). During the years 1878–1879, the University of Genoa hired Edoardo Maragliano to
teach general pathology. Then, in 1880, he
became professor of internal medicine at the
University of Cagliari. A year later, the University of Genoa appointed Maragliano as full professor of internal medicine and named him
director of the Medical Clinic of Genoa, a post
he held until 1924. After the death of the anato-
mist Professor Luigi Ageno (1822–1884),
Maragliano became the Dean of the Medical
Faculty until 1893. In 1907, he was elected Rec-
tor of the University of Genoa, a position he held
for 10 years. In this period, he enlarged the
ancient "Pammatone" Hospital in Genoa through
the construction of new buildings, which
constituted the nucleus of the modern Hospital
of San Martino in Genoa (Massini 1907;
Costantini 1950).

Maragliano's scientific activity was intense
and varied. As a passionate researcher in all
fields of medicine, he contributed to the understanding of the pathophysiology of hemic
diseases, nephritis, hepatopathies, brain
disorders, metabolic disorders and, above all,
infectious diseases (pneumonia, cholera, typhus,
flu; Massini 1907; Costantini 1950; Maragliano
1889). The activities of all researchers in the
Medical Clinic and laboratories directed by

Maragliano are attested by the 2,453 scientific works published under his supervision (Cardinale 1995). In addition to the scientific institutions which he contributed to developing (the Royal Medical Academy of Genoa in 1885, and the Italian Society of Internal Medicine, founded in 1887 with Guido Baccelli and Arnaldo Cantani and directed by Maragliano himself from 1919 until his death), some initiatives were undertaken not only for clinical and research purposes, but as a result of his profound social awareness. In 1872, Maragliano and other members of civil society founded a sea rescue association, which still operates today; this led to the institution of maritime lookout posts positioned along the Ligurian coast (Massini 1907). Then, in 1896, Maragliano created the first Sanatorium in Italy, providing clinical and social support for patients and their families (Benedicenti 1940). From 1900 until his death, as a member of the Senate of the Kingdom of Italy, he discussed the budgets of the ministries that dealt with social, economic and sanitary questions, without neglecting his duties towards his patients and students; to do so, he commuted overnight by train between Rome and Genoa.

In 1900, an institution for study of infectious disease started research into tuberculosis, under the supervision of Maragliano. His sensitivity towards the pathogenesis, therapy and prevention of tuberculosis first developed during his training as a student of medicine in Naples, and was probably heightened by the death, from pulmonary tuberculosis, of his brother, Dario (1854–1889), a psychiatrist (Costantini 1950; Massini 1907).

After the discovery of *M. tuberculosis* as the causative agent of tuberculosis, the scientific community started to search for therapy (Barberis et al. 2017). Nevertheless, because of the inconclusive results of the first studies, a negative attitude arose towards the possibility of understanding the pathogenesis of the disease and establishing any method of activating immunity (Salvioli 1956; Pizzini 1896). Only the technique of artificial pneumothorax designed by Carlo Forlanini (1847–1918) in 1883, seemed to be the best therapy for pulmonary tuberculosis in terms of effectiveness (Premuda 1975). In 1890, however, Maragliano and his collaborators began experimental and clinical studies to demonstrate the existence of innate immunity and specific immunity against *M. tuberculosis* (Maragliano 1904). Indeed, Maragliano was the first scientist to affirm that the fight against tuberculosis should be based on specific immunization.

After 5 years of research, at the Second Congress of the French Society of Internal Medicine held in Bordeaux on 12th August 1895, Maragliano was able to announce the existence of a tubercular antitoxin in infected animals (dogs, asses and horses), and the consequent use of animal serum for serotherapy (Pizzini 1896; Maragliano 1903; Salvioli 1956). In his following studies, Maragliano demonstrated that the "tubercular antitoxin" included components of the innate and specific immune systems, produced by leukocytes. All these components were obtained experimentally by infecting animals with attenuated or inactivated bacteria or bacterial derivatives, which always originated from the same defensive process (Maragliano 1907). He observed that infecting a human or animal organism with bacteria elicited greater synthesis of antibodies than did infection with bacterial derivatives; the latter displayed a great ability to induce components of the innate immune system. Pathological conditions, such as concomitant diseases and undernourishment, and physiological conditions, such as pregnancy, puerperium and feeding, were seen to compromise the efficacy of immune systems, especially the innate system. According to Maragliano, passive immunization achieved through oral or serotherapy lasted between 2 and 3 years. Although scientists such as Emil Adolf von Behring (1854–1917) claimed that the only effective vaccine against tuberculosis had to contain live bacteria, Maragliano and his collaborators avoided using live microorganisms, in order to obtain an innocuous vaccine, without bacterial activity. They therefore, proposed the inoculation of an inactivated vaccine (Maragliano 1904).

After 1895, Maragliano reported the results of his studies at medical congresses: in Madrid and Padua in 1903, in Philadelphia in 1904, and in Lyons and The Hague in 1906 (Massini 1907; Salvioli 1956). The initial reaction of the scientific community was one of scepticism. This was followed by harsh criticism and controversy, particularly with regard to the use of an inactivated vaccine, even though scientists such as Behring and Koch admitted the existence of an immune response against M. tuberculosis in their studies subsequent to Maragliano's statement in Bordeaux (Pizzini 1896).

Maragliano's serum was tested in various parts of Italy, not least because serotherapy had proved successful in treating other infectious diseases, such as diphtheria (Tognotti 2012). However, serotherapy displayed clinical efficacy only in patients who were not undernourished or suffering from concomitant infections or severe pulmonary impairment with extensive destruction of the pulmonary parenchyma. In spite of the limited effectiveness of serotherapy, combining serotherapy with improvements in patients' environmental and nutritional conditions would, according to Maragliano, "provide the energy required to fight successfully against tuberculosis", hence the social and clinical need for sanatoriums (Maragliano 1904). In 1904, Maragliano's study of the first inactivated vaccine aroused hope for a broad application of this new preventive measure in homes and workplaces where tuberculosis was widespread (Maragliano 1907). The possibility of vaccinating people against tuberculosis was disparaged by the scientific community until 1921, when Albert Calmette (1863–1933) developed and proposed an attenuated vaccine. This discovery was sustained and promoted by French public institutions and marked the lasting success of the Bacillus Calmette–Guèrin (BCG) vaccine. During the 6th Conference of the International Union against Tuberculosis, held in Rome in 1928, Calmette acknowledged that Maragliano and his school had been the first to study and promote vaccination against tuberculosis. Nevertheless, Calmette himself disapproved of the use of inactivated vaccine on the grounds that inactivated microorganisms did not have any immunogenic potency. However, during the 7th International Conference, which was held in Oslo in 1930, Maragliano demonstrated that criticism regarding the ineffectiveness of inactivated vaccine was unfounded, since immunity could be elicited by antigens of both inactivated and attenuated bacteria. Indeed, he firmly believed that any type of vaccine was able to elicit immunity: "Vaccinate, vaccinate! The preventive use of antigens obtained from attenuated or inactivated microorganisms is greatly helpful in any case" (Martini and Paluan 2017; Costantini 1950; Salvioli 1956; Urizio 1940).

Only in 1951 Maragliano's discoveries were confirmed by the scientific community, when, at the Academy of Sciences in Paris, Robert Debré (1882–1978) and Gaston Ramon (1886–1963) reported that inactivated vaccines elicited immunity after the same time-lag and at the same intensity as BCG vaccines. Before this declaration, inactivated vaccines prepared by Maragliano or Giovanni Petragnani (1893–1969) had successfully been used in northern Italy (Salvioli 1956).

In order to investigate diseases from a pathophysiological point of view, Maragliano provided the laboratories of his clinic with equipment for microbiology, serology, hematology, biochemistry and, for the first time in Italy, radiology. Indeed, after the discovery of X rays in 1895 by Wilhelm Conrad Roentgen (1845–1923), Maragliano immediately realized that this new radiation could be applied to clinical cases. Thus, in 1896, he installed the first radiological equipment in his clinic, thereby promoting the development of radiology as an independent discipline. Although technologies were very limited (because of the low power of bobbins and the low performance of X-ray tubes), Maragliano foresaw an increasing use of this technique in clinical practice and, in December 1896, made his first radioscopic observations on patients in his clinic. An associate of Maragliano's, the physician Vincenzo Sciolla, was designated to test the instrumentation; but after his premature death in March 1897, the radiology laboratory was entrusted to the medical doctor Marco Sciallero, until 1909, when Sciallero withdrew owing to radiation injuries (Tedeschi 1907; Costantini 1950; Cardinale 1995; Massini 1907). From the age of 19, Maragliano's son Vittorio attended the radiology laboratory, where he studied the displacement of his own heart on adopting various positions of the body. In 1913, Vittorio became the first full professor of radiology in Italy and, from 1919 to 1922, the second president of the Italian Society of Radiology. He died of radiation injuries in 1944 (Cardinale 1995; Costantini 1950; Massini 1907).

Edoardo Maragliano asserted that "The first duty of a physician is to lead young colleagues to a correct diagnosis, a firm prognosis and a useful therapy, by applying all the achievements of science at the patient's bedside". For this reason, during his lessons, he submitted more than three clinical cases to his students, in the presence of patients themselves into the lecture theatre. Maragliano clearly and simply explained and discussed with his students the most frequent diseases a physician would be likely to encounter in his career (nephritis, pneumonia, rheumatic valvular disease, typhus, infectious gastroenteritis; Costantini 1950, Benedicenti 1940). On account of his medical experience and scientific knowledge, Maragliano oversaw, together with his mentor, Professor Arnaldo Cantani, the drafting of the 24 volumes called "Italian Treatise of Medical Pathology and Therapy" (Urizio 1940).

In 1915, after the Kingdom of Italy had declared war on the Austro-Hungarian Empire, Maragliano enlisted as a volunteer in the Italian Army, with the rank of Major General. During the war, he organized medical services and centres for the diagnosis of tuberculosis both behind the lines and on the battlefield. Moreover, together with Luigi Lucatello, he directed the Medical Clinic at the Military University of Padua (Costantini 1950).

Even after leaving his post as professor of internal medicine in 1924, Maragliano remained active and kept his fervor for research into tuberculosis, creating in Genoa the first Italian school of specialization for the study of tuberculosis, and gathering a large group of young collaborators (Urizio 1940; Benedicenti 1940). He spent the last years of his life teaching in this school, in order to create a new specialization in medicine, the purpose of which was to give physicians specific training in the diagnosis, therapy and prevention of tuberculosis (Argentina 1931). On 10 March 1940, at the age of 90 years, he died at his home in Genoa (Benedicenti 1940).

Edoardo Maragliano lived in a period rich in scientific discoveries in medicine both in Italy and in Europe, surrounded by masters such as Guido Baccelli, Augusto Murri, Achille De Giovanni, Antonio Cardarelli, Robert Koch, Rudolf Virchow (1821–1902), Claude Bernard (1813–1878) and Charles Jacques Bouchard (1837–1915). Maragliano was able to keep abreast of medical advances and hone his innate talents in Naples, where he followed Professor Cantani in the medical clinic and laboratories of the university. Moreover, in the poorest districts of the city, he braved disease and danger in order to attend patients affected by cholera. A follower of positivism and the experimental method, based on clinical observation and research in the laboratory, he practiced and taught medicine with enthusiasm and dedication at the bedsides of patients and in the university, until his death. Maragliano is considered the true founder of phthisiology in Italy – the whole set of concepts and actions aimed at protecting mankind against tuberculosis, above all through preventive means, such as vaccination.

Acknowledgements and Note We thank Bernard Patrick and Giorgia Nicholà Monaco for linguistic revision.

References

Argentina G (1931) L'inizio clinico della infezione tubercolare: accertamento clinico, cenni di radiodiagnostica e di tecnica di laboratorio. Tipografia Pellegrini, Pisa, pp II–V

Barberis I, Bragazzi NL, Galluzzo L et al (2017) The history of tuberculosis: from the first historical records to the isolation of Koch's bacillus. J Prev Med 58:E9–E12

Benedicenti A (1940) Edoardo Maragliano: commemorazione letta alla Società di Scienze e Lettere di Genova. In: Atti della Società di Scienze e Lettere di Genova. Premiata Tipografia Successori Fratelli Fusi, Pavia, pp 16–24

Borghi L (2015) Il medico di Roma. Vita, morte e miracoli di Guido Baccelli (1830–1916). Armando Editore, Rome, pp 234–237

Cardinale AE (1995) Radiologic pantheon. Edoardo Maragliano and the Italian forerunners of the specialty. Radiol Med 89:573–577

Costantini G (1950) Edoardo Maragliano. Riforma Med 64:141–145

Maragliano E (1889) Rimedi nuovi e nuovi metodi di cura. Manuale di terapia clinica moderna ad uso dei medici e degli studenti. Vallardi, Milan, vol. 2, pp 3–23 and 295–317

Maragliano E (1903) La lotta e la immunizzazione dell'organismo contro la tubercolosi. In: Conferenza tenuta al Congresso Internazionale Medico di Madrid. Vallardi, Milan, pp 6–13

Maragliano E (1904) La terapia specifica della tubercolosi e la sua vaccinazione. Vallardi, Milan, pp 4–34

Maragliano E (1907) Sullo stato attuale della terapia specifica della tubercolosi. Vallardi, Milan, pp 4–41

Martini M, Paluan F (2017) Edoardo Maragliano (1849–1940): the unfortunate fate of a real pioneer in the fight against tuberculosis. Tuberculosis 106:123

Massini G (1907) Edoardo Maragliano nella vita e nella scienza. In: Rivista Popolare d'Igiene e Medicina. Fratelli Vaccarezza, Genoa, pp 11–15

Pizzini L (1896) Il Prof. Edoardo Maragliano e la sieroterapia della tisi, vol IV. Istituto Italiano di arti grafiche, Bergamo, pp 136–143

Premuda L (1975) Storia della Medicina. CEDAM, Padua, pp 242–244

Salvioli G (1956) In memoria di Edoardo Maragliano. Annali. Sanita Pubblica 17:887–893

Tedeschi E (1907) Le onoranze al prof. Edoardo Maragliano pel suo 25° anno d'insegnamento. In: Cronaca della Clinica Medica di Genova. Stabilimento Tipografico G. Testa, Biella, pp 201–224

Tognotti E (2012) Il morbo lento, la tisi nell'Italia dell'Ottocento. F. Angeli editore, Sassari, pp 157–161

Urizio L (1940) Commemorazione del Senatore Edoardo Maragliano. In: Bollettino dell'Associazione medica triestina. Tipografia del Lloyd triestino, Trieste, pp 250–252

Adv Exp Med Biol - Advances in Microbiology, Infectious Diseases and Public Health (2018) 9: 101–110
DOI 10.1007/5584_2017_134
© Springer International Publishing AG 2017
Published online: 27 December 2017

The Efficacy of Tetrasodium EDTA on Biofilms

S.L. Percival and A-M. Salisbury

Abstract

The aetiology of delayed wound healing characteristic of a chronic wound is relatively unknown but is thought to be due to a combination of the patient's underlying pathophysiology and external factors including infection and biofilm formation. The invasion of the wound by the hosts' resident microbiome and exogenous microorganisms can lead to biofilm formation. Biofilms have increased tolerance to antimicrobial interventions and constitute a concern to chronic wound healing. Consequently, anti-biofilm technologies with proven efficacy in areas outside of wound care need evaluation to determine whether their efficacy could be relevant to the control of biofilms in wounds. The aim of this study was to assess the anti-biofilm capabilities of tetrasodium EDTA (t-EDTA) as a stand-alone liquid and when incorporated in low concentrations into wound dressing prototypes. Results demonstrated that a low concentration of t-EDTA (4%) solution was able to kill *Staphylococcus aureus*, methicillin-resistant *S. aureus* (MRSA), *S. epidermidis*, *Pseudomonas aeruginosa* and *Enterococcus faecalis* within in vitro biofilms after a 24-h contact time. The incorporation of low levels of t-EDTA into prototype fibrous wound dressings resulted in a 3-log reduction of bacteria demonstrating its microbicidal ability. Furthermore, hydrogels incorporating only a 0.2% concentration of t-EDTA (at preservative levels) caused a small reduction in biofilm. In conclusion, these studies show that t-EDTA as a stand-alone agent is an effective anti-biofilm agent in vitro. We have demonstrated that t-EDTA is compatible with numerous wound dressing platforms. EDTA could provide an essential tool to manage biofilm-related infections and should be considered as an anti-biofilm agent alone or in combination with other antimicrobials or technologies for increased antimicrobial performance in recalcitrant wounds.

Keywords

Antimicrobial · Biofilm · EDTA · Infection · Wound Dressing · Wounds

1 Introduction

The formation of biofilms in medical devices or acute and chronic wounds is of great importance (Francolini and Donelli 2010, Percival et al. 2012, 2015; Percival 2017). Biofilms have been identified in chronic wounds using microscopy methods, and it has been demonstrated that these wound biofilms are polymicrobial in nature (Wolcott et al. 2013). The clinical significance of biofilms relates to their

S. L. Percival (✉) and A.-M. Salisbury
Centre of Excellence in Biofilm Science and Technologies (CEBST), 5D Health Protection Group Ltd, Liverpool Bio-innovation Hub, Liverpool, UK
e-mail: Steven.Percival@5Dhpg.com

increased tolerance and recalcitrance to antimicrobial interventions. Biofilms are microorganisms that attach to biotic and abiotic surfaces, or themselves forming aggregates, and produce an encasing polymeric substance known as exopolymeric substance (EPS) (Wingender et al. 2012). The presence of metal ions in the biofilm is significant to its development, sustainability and maintenance (Donlan 2002). Such divalent cations such as magnesium and calcium have been shown to cross-link with the polymer strands in EPS to provide greater binding force in a developed biofilm (Donlan 2002). Many studies have shown that the treatment of biofilms with metal ion chelators such as ethylenediaminetetraacetic acid (EDTA) helps to reduce biofilm formation or aid the removal of established biofilm (Devine et al. 2007; Finnegan and Percival 2014; Kite et al. 2005; Moreau-Marquis et al. 2009). Iron permits bacterial differentiation and essential biofilm growth, and therefore biofilm formation can be supressed by limiting available iron (Weinberg 2004; Che et al. 2009). The use of EDTA to control biofilm formation has also been well documented and has been shown to outperform heparin in the prevention of biofilm growth in catheter-related blood stream infections (Percival et al. 2005; Kite et al. 2004; Raad et al. 2008). Interestingly, the conventional use of heparin has been shown to stimulate *Staphylococcus aureus* biofilm formation (Shanks et al. 2005). Despite the suggested use of using EDTA in wound care dressings documented in 2001 (Kite and Hatton 2001), and in combination with other agents (Kite et al. 2005; Percival et al. 2017), its application has only relatively recently been utilised in wound dressings despite extensive safety and toxicology profiles (Anon 2004).

EDTA when combined with sodium exists in many different forms and at different pH ranges and different concentrations in medical devices and has differences in their antimicrobial and anti-biofilm properties (Kite and Hatton 2001; Kite et al. 2005). The use of tri- (tri-EDTA, pH 7 to 9) and tetrasodium EDTA (t-EDTA, above pH 9) as standalone antimicrobial and anti-biofilm agents, or as synergistic agents to other antimicrobials, is being explored and has been reviewed elsewhere (Finnegan and Percival 2014). Both tri-EDTA and t-EDTA have shown anti-biofilm capabilities in part due to the chelation of metal ions including calcium, zinc, magnesium and iron, which affects the stability of the biofilm and can also potentiate and sensitise the cell walls of bacteria (Banin et al. 2006).

Therefore, the aim of this paper was to assess the effect of t-EDTA only, as we have shown it has very similar antimicrobial and anti-biofilm performance to tri-EDTA (data not shown), on in vitro biofilm models and, furthermore, whether the incorporation of t-EDTA into various wound dressing platforms at both preservative (0.2%) and potential therapeutic concentrations (2–8%) would be effective on in vitro biofilms.

2 Methods

2.1 Microorganisms

P. aeruginosa NCTC 10662, *Staphylococcus aureus* ATCC 25923, *S. aureus* ATCC 29213, methicillin-resistant *S. aureus* (MRSA) ATCC BAA-43, *S. epidermidis* ATCC 35984 and *Enterococcus faecalis* ATCC 29212 were tested in the Minimum Biofilm Eradication Concentration (MBEC) model. *P. aeruginosa* NCTC 10662 and *S. aureus* ATCC 29213 were further tested in the Centers for Disease Control and Prevention (CDC) bioreactor and filter biofilm models, and *E. faecalis* ATCC 29212 was tested in the confocal study.

2.2 Chemicals

Tetrasodium EDTA was obtained from Sigma-Aldrich (UK).

2.3 Wound Dressings

Multisorb (BSN Medical, UK) absorbent dressings (fibrous gauze dressing) were used as basic gauze dressings to incorporate t-EDTA formulations. Briefly, 5 cm × 5 cm pieces of fibrous dressings were soaked in 2% and 4% t-EDTA for 1 h, weighed and allowed to dry in a sterile environment. T-EDTA was incorporated into amorphous hydrogels with a final concentration of t-EDTA of 0.2%.

2.4 Direct Contact Antimicrobial Test

The direct contact method was adapted from British Standards BS EN 16756 Antimicrobial Wound Dressings - Requirements And Test Methods (Section H.6). Control dressings and t-EDTA-incorporated wound dressings were aseptically cut to 2.5 cm × 2.5 cm and placed into 50 mL sterile centrifuge tubes. An overnight culture of *S. aureus* and *P. aeruginosa* was diluted to 10^8 CFU/mL, and 500μL was added of each microorganism to the dressings. Dressings were incubated for 24 h at 37 °C +/− 2 °C before adding 10 mL of Tryptone Soya Broth (TSB) (Oxoid, UK). All tubes were vortexed and sonicated to remove attached bacteria from the wound dressings before performing serial dilutions and determining total viable counts (TVCs).

2.5 Biofilm Models

2.5.1 Minimum Biofilm Eradication Concentration (MBEC) Model

The method for the use of the MBEC biofilm model was adapted from ASTM E2799. Briefly, an overnight suspension of microorganisms was diluted to 1×10^5 CFU/ml before inoculating the wells of the 96-well plate. The 96-peg lid was added to the plate and incubated for either 24 h (24-h biofilm) or 48 h (48-h biofilm) at 37 °C+/−2 °C and agitated at 125 rpm in humidified conditions. Biofilms were washed in sterile 0.85% sodium chloride (Sigma, UK) solution and placed into a new plate containing 150μl of 4% t-EDTA. Biofilms or the plates were incubated for 24 h. All pegs were washed and sonicated for 30 min into sterile 0.85% saline solution before serially diluting for total viable counts (TVCs).

2.5.2 CDC Bioreactor Biofilm Model

A modified ASTM E2871–13, Standard Test Method for Evaluating Disinfectant Efficacy Against *Pseudomonas aeruginosa* Biofilm Grown In CDC Biofilm Reactor Using Single Tube Method, was used. Briefly, Tryptone Soya Broth was inoculated with either *S. aureus* ATCC 29213 or *P. aeruginosa* NCTC 10662 to a concentration of 1×10^8 CFU/ml, which was determined by optical density (at 600 nm) and total viable counts. Each CDC bioreactor (BioSurface Technologies, USA) contains eight polypropylene rods designed to hold three coupons. In this experiment, polycarbonate coupons were used. The CDC reactor was sterilised before aseptically adding 300 ml of sterile TSB through the inoculation port. Following this, 1 ml of the previously prepared 10^8 CFU/ml inoculum was then added to the reactor. The reactor was placed on a magnetic stir plate, and the rotation speed was set to 125 ± 5 rpm. The CDC reactor was operated in batch mode at room temperature (21 ± 2 °C) for 48 h (48-h biofilms). Following incubation, the rods containing the polycarbonate coupons were removed and rinsed in sterile 0.85% sodium chloride solution to remove planktonic cells. Each coupon was released from the rods into individual sterile 50 mL centrifuge tubes. For the testing of liquid t-EDTA (stand-alone), each coupon was placed into 3 mL of 4% t-EDTA. For the testing of hydrogels, each coupon was treated with 3 g of either hydrogel (control) or hydrogel containing 0.2% t-EDTA. Treatments of the coupons were performed in triplicate. The coupons were then incubated at 37 °C for 24 h. Following incubation, 27ml of sterile distilled water was added to each tube and sonicated for 30 min, followed by mixing for 10 s using a vortex to ensure the bacterial cells were in suspension. The disaggregated biofilm samples were sampled and serially diluted for bacterial enumeration. Biofilm density was calculated Log_{10} density for each coupon. The Log_{10} density of each coupon was subtracted from the Log_{10} density of the untreated control coupon to determine the Log_{10} reduction value of each treated biofilm.

2.5.3 Bespoke Biofilm Filter Model

S. aureus or *P. aeruginosa* was cultured overnight in TSB at 37 °C before diluting to 1×10^8 CFU/ml. Twenty microliters of the bacterial suspension was added to a sterile 0.2μm filter disc that was placed on Tryptone Soya Agar (TSA) (Oxoid, UK). Bacteria were incubated at 37 °C+/ −2 °C in humidified conditions for 48 h (48-h biofilm). Following incubation, 2 cm × 2 cm gauze (control), gauze with 2% t-EDTA or gauze with 4% t-EDTA was hydrated with 600µl sterile distilled water and placed on top of the biofilm. Biofilms were incubated for a further 24 h before removing and discarding the dressings. The filter containing the biofilm was then placed in 10ml of sterile distilled water, sonicated on full power for 30 min and vortexed for 10 s to remove the biofilm. The resulting suspension was used to perform total viable counts (TVCs) using serial dilution and plating.

2.6   Confocal Analysis

Overnight cultures of bacteria were diluted to 10^6 CFU/mL before adding 400µl of the inoculum to the wells of a Lab-Tek™ glass chamber slide. The chamber slides were incubated at 37°C ± 2°C in humidified conditions for 48 h. Following incubation, 48-h biofilms were washed twice in 0.85% sodium chloride solution to remove planktonic cells. Biofilms were then stained using the LIVE/DEAD BacLight stain kit (Thermo-Fisher, UK) for 20 min at room temperature. Biofilms were then washed once in 0.85% sodium chloride solution and images were taken (T0) using the Zeiss 710 confocal laser scanning microscope. Biofilms were then treated with 400µl of t-EDTA at 0.2%, 2% and 4%. Control biofilms were in TSB only. Biofilms were incubated for 24 h at 37°C ± 2°C in humidified conditions. Images were taken at T24 (24-h contact time) hours using the Zeiss 710 confocal laser scanning microscope. Images were processed using FIJI software.

2.7   Cytotoxicity

2.7.1   Indirect Cytotoxicity (Lysis Index)

A 3% solution of Agar (suitable for cell culture) (Sigma, UK) was made and autoclaved for 15 min at 121 °C. The autoclaved agar was put into a 45 °C

water bath and allowed to cool to 45 °C. DMEM was warmed to 45 °C in a water bath. Equal volumes of the DMEM and 3% agar were mixed and allowed to cool to approximately 39 °C. The medium from all acceptable cultures was removed and replaced with 2.0 mL of agar medium. The cultures were placed on a flat surface to solidify at room temperature. Sterile, 10 mm blank antibiotic susceptibility discs (Sigma, UK) were saturated in test solution or control solution before being placed in each dish in contact with the agar surface. Triplicate cultures for each test material and both positive and negative controls were performed. All cultures were incubated for 24 ± 1 h. The outline of the specimen was marked on the bottom of the culture dish with a permanent marker, and then the specimen was removed. Two millilitres of 0.01% neutral red solution was added to each dish and incubated for 1 h. Following incubation, the neutral red solution was removed and each culture was examined microscopically under and around each control and test specimen. The cell culture was deemed to show a cytotoxic effect if microscopic examination revealed malformation, degeneration, sloughing or lysis of the cells within the zone or a moderate to severe reduction in cell layer density. The lysis index (Table 1) measures the number of cells affected within the zone of toxicity.

Table 1 Qualitative lysis description

Lysis index	Description of zone	Reactivity
0	Discrete intracytoplasmic granules, no cell lysis, no reduction of cell growth	None
1	Not more than 20% of zone shows rounded cells, loosely attached and without intracytoplasmic granules or show changes in morphology	Slight
2	Not more than 50% of the cells are round, devoid of intracytoplasmic granules, no extensive cell lysis; not more than 50% growth inhibition observable	Mild
3	Not more than 70% of the cell layers contain rounded cells or are lysed; cell layers not completely destroyed, but more than 50% growth inhibition observable	Moderate
4	Nearly complete destruction of the cell layers	Severe

2.8   Statistics

Statistical analysis was performed using GraphPad Prism 7 software. Statistical comparisons were performed using the two-way ANOVA. Results were interpreted as significant if the p value was ≤ 0.05.

3   Results

3.1   Effectiveness of T-EDTA on Biofilms After Short and Long Contact Times

The treatment of 24-h and 72-h *S. aureus* and *P. aeruginosa* biofilms with a 4% t-EDTA solution for short contact times (1, 5 and 15 min) in the CDC bioreactor model resulted in a reduction in biofilms from 54.1% to 99.6% at the various short contact times (Figs. 1 and 2). Treatment of 24-h *S. aureus* biofilms for 1- and 15-min contact times demonstrated a significant reduction in cell density ($p = <0.04$). A significant reduction in cell density following treatment of 24- and 72-h *P. aeruginosa* biofilms was found at all contact times ($p = <0.0001$). The efficacy of 4% t-EDTA was demonstrated when exposing 24- and 48-h biofilms to t-EDTA for 24 h, whereby treatment caused 100% kill of *S. aureus*, MRSA, *S. epidermidis*, *E. faecalis* and *P. aeruginosa* biofilms (Fig. 3).

3.2   Effectiveness of Tetrasodium EDTA on Biofilms when Incorporated into Fibrous, Gauze and Hydrogel Prototype Platforms

The incorporation of t-EDTA at varying concentrations into fibrous prototype dressings was evaluated using a direct contact method (BS EN 16756) against *S. aureus* and *P. aeruginosa*. To test the effectiveness of t-EDTA-incorporated platforms against biofilms, 20 mg per gram (mg/g) (2%) and 40 mg/g (4%) of t-EDTA were added to the gauze and its efficacy was tested against a bespoke filter biofilm model. Results showed the gauze alone resulted in a small decline in cell density, which is likely due to mechanical disruption of the biofilm. Results showed 24-h treatment of *S. aureus* and *P. aeruginosa* biofilms caused moderate log reductions with gauze containing 4% t-EDTA causing greater log reductions than gauze containing 2% t-EDTA (Fig. 4). Incorporating 4% t-EDTA into the dressing resulted in a significant reduction of cell density against both strains in comparison to the control ($p = <0.002$) as well as the gauze only ($p = <0.04$) (Fig. 4).

Hydrogel only and hydrogel plus 0.2% t-EDTA resulted in a significant reduction of bacterial cell density against 48-h biofilms of *S. aureus,* MRSA, *P. aeruginosa, S. epidermidis* and *E. faecalis* ($p = <0.035$) with the exception

Fig. 1 Mean log density of 24- and 72-h *S. aureus* biofilms following short exposure to 4% t-EDTA. *S. aureus* biofilms were created using the CDC biofilm bioreactor and treated with 4% t-EDTA for 1, 5 and 15 min. Testing was done in triplicate. Error bars represent the standard deviation. A significant reduction in bacterial cell density compared to a control was found with some treatment groups at 24 h (* $p = <0.04$)

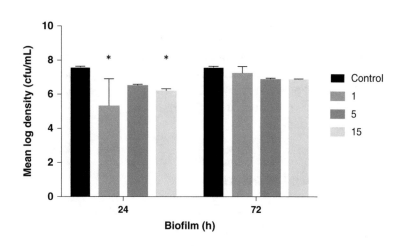

Fig. 2 Mean log density of 24- and 72-h *P. aeruginosa* biofilms following short exposure to 4% t-EDTA. *P. aeruginosa* biofilms were created using the CDC biofilm bioreactor and treated with 4% t-EDTA for 1, 5 and 15 min. Testing was done in triplicate. Error bars represent the standard deviation. A significant reduction in the bacterial cell density of 24- and 72-h biofilms compared to a control was found with all treatment groups (* $p = <0.0001$).

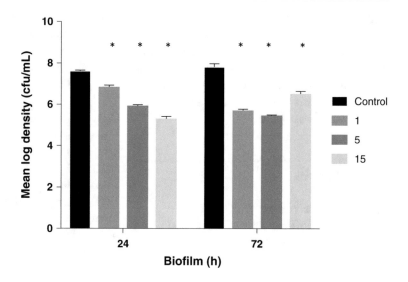

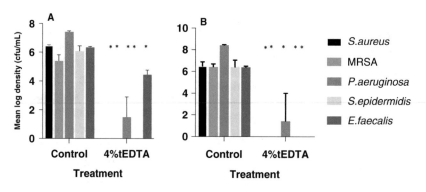

Fig. 3 MBEC results for treatment with t-EDTA. 24- (A) and 48- (B) hour biofilms were formed using MBEC plate method and treated with 4% t-EDTA for 24 h. Testing was done in triplicate. Error bars represent the standard deviation. A significant reduction in biofilm cell density was found following treatment with 4% t-EDTA in comparison to a control group with all 5 strains (* $p = <0.0006$)

of hydrogel only against *S. epidermidis*. The reduction of cell density with hydrogel only is likely due to disruption of the biofilm upon application. A significant reduction in cell density was found with hydrogel plus 0.2% t-EDTA with all 5 strains when compared to hydrogel alone ($p = <0.013$) (Fig. 5).

3.3 Confocal Microscopy

Treatment of the *S. epidermidis* biofilm with all concentrations of t-EDTA resulted in a reduction in viability and changes in biofilm architecture.

Treatment of *S. aureus*, MRSA and *E. faecalis* (Fig. 6) with t-EDTA for 24 h reduced the viability of attached cells. The reduction in biofilm formation appeared to be dose-dependent based in all confocal images analysed.

3.4 Cytotoxicity of Tetrasodium EDTA

In the lysis index, it was shown that 2%, 4% and 8% of t-EDTA were non-cytotoxic, with grades 2, 1 and 2, respectively (Table 2).

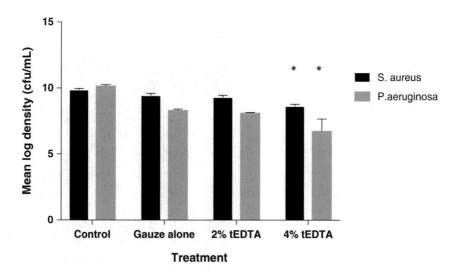

Fig. 4 The antimicrobial activity of t-EDTA-incorporated into fibrous dressing against *S. aureus* and *P. aeruginosa* using the direct contact method. Testing was done in triplicate. Error bars represent the standard deviation. Results showed that the gauze alone had an effect on the biofilm density, so t-EDTA treatment groups were compared to the gauze alone to determine a significant reduction in cfu/mL (* $p = <0.04$)

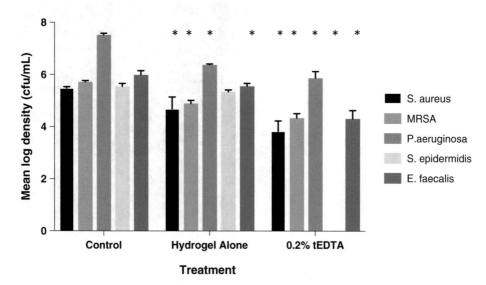

Fig. 5 Mean log density of 48-h biofilms following treatment with hydrogel alone and hydrogel with 0.2% t-EDTA in the CDC model. Biofilms (48 h) were exposed to treatment for 24 h at $37\ °C \pm 2\ °C$. Samples were run in triplicate. Error bars represent standard deviation. A significant log reduction in biofilm density was found in both treatment groups compared to the control (* $p = <0.035$)

4 Discussion

Within this study we have shown that tetrasodium EDTA, as a solution, and within certain wound prototype dressing formats helps to remove established biofilm and kill microorganisms that had been growing within the biofilm. Interestingly, the ability of 4% t-EDTA to remove established biofilm varied in the different models utilised in this study. The

Fig. 6 Treatment of
E. faecalis biofilms with
tetrasodium EDTA.
Biofilms were treated with
0.2%, 2% and 4%
tetrasodium EDTA for
24 h. Control biofilm was in
Tryptone Soya Broth.
Green fluorescence
represents live cells. Red
fluorescence represents
dead cells. Images were
taken using Zeiss confocal
laser scanning microscope
at x20 magnification with
Z-stacks. Scale bar
represents 50μm

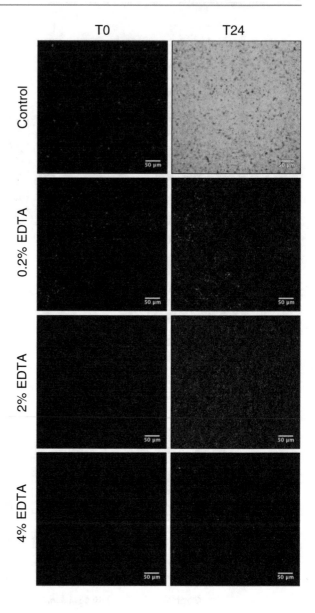

MBEC model showed 24-h treatment with 4% t-EDTA is effective in significantly reducing 24- and 48-h biofilm cell density to below detectable levels against the majority of 5 strains tested. In comparison, although 4% t-EDTA reduced the bacterial cell density of biofilms in the CDC bioreactor, this was to a lesser extent. The difference observed between the 2 models could be due to a higher bacterial cell density being achieved in the CDC bioreactor or due to differences in biofilm formation in each model.

For example, the plates used in the MBEC are polystyrene, whereas the coupons used in the CDC bioreactor model are polypropylene. Different materials can affect the ability of bacterial cells to attach to a surface and form biofilms. The difference in results from the MBEC and CDC bioreactor experiments shows the importance of utilising different biofilm models when determining the anti-biofilm properties of a test agent. Differences in the anti-biofilm efficacy of the hydrogel test dressing and tetrasodium EDTA

Table 2 Zone index and lysis index for indirect cytotoxicity test

Sample	Lysis index	Interpretation
DMEM	0	Non-cytotoxic
Water	0	Non-cytotoxic
Phenol	3	Cytotoxic
EDTA 2%	2	Non-cytotoxic
EDTA 4%	1	Non-cytotoxic
EDTA 8%	2	Non-cytotoxic

Zone index measures the clear zone in which cells do not stain with neutral red. The lysis index measures the number of cells affected within the zone of toxicity. All samples were tested in triplicate

alone (2% and 4%) were due to a lower concentration of tetrasodium EDTA in the hydrogel formulation (0.2%). However, low levels of t-EDTA were added at low concentration to see if there was a preservative effect in the hydrogel. Further work continues at higher concentrations of tetrasodium EDTA.

The incorporation of tetrasodium EDTA into fibrous prototype platforms resulted in at least a 3-log reduction in the direct contact method demonstrating a microbicidal effect according to international guidelines. It has been previously shown that the use of liquid-only tetrasodium EDTA ((40 mg/mL, 4%) and (20 mg/ml, 2%)) as a catheter lock solution is effective in the reduction of biofilm in a catheter model (Percival et al. 2005; Kite et al. 2004). Similarly, tetrasodium EDTA has been reported to help with the partial removal of biofilm from polymethyl methacrylate (PMMA) and toothbrushes (Devine et al. 2007). Other studies have also assessed the effects of EDTA in biofilm prevention and control, which resulted in positive biofilm prevention in a dose-dependent manner; however the reduction of pre-formed biofilms was minimal with only a 31% reduction recorded (Ramage et al. 2007). EDTA in conjunction with other metals and solutions has demonstrated significant efficacy on biofilms. Banin and colleagues also determined that EDTA in Tris buffer was a thousand times more effective in killing *P. aeruginosa* biofilm

than gentamicin alone. Furthermore, the excess of divalent cations, magnesium, calcium and iron, protected the biofilm from anti-biofilm EDTA activity, supporting the importance of these divalent irons in the protection of the biofilm (Banin et al. 2006).

Within this study we have demonstrated that tetrasodium EDTA showed moderate to no cytotoxicity in the indirect contact method. Other studies support this finding whereby neutral and alkali tetrasodium EDTA cause moderate to severe cytotoxicity in L929 cells in a dose-dependent manner (Koulaouzidou et al. 1999). EDTA is a strong organic acid that is approximately 1,000 times stronger than acetic acid and does not appear to occur naturally. Toxicological data shows that the oral lethal doses of disodium EDTA and trisodium EDTA salts in rat are 2000 mg/kg and 2,150 mg/kg, respectively (Anon 2004; Kimmel 1977; Schardein et al. 1981). Extensive reviews of EDTA have determined a low toxicity profile for humans and no risk to human health (Grundler et al. 2005).

Overall we have shown that it is possible to incorporate tetrasodium EDTA into various wound dressing prototype materials such as gauze (fibrous material) and hydrogels. The use of tetrasodium EDTA as an effective microbicidal, antimicrobial and anti-biofilm agent has been demonstrated in this study indicating that concentration levels and the concentrations incorporated into wound dressings, and the amounts eluting from the dressing, are important and can affect anti-biofilm ability. As is the case for all microbicidal technologies, the efficacy of t-EDTA on planktonic microbes, sessile microbes and biofilms is dose-dependent. Consequently, the use of tetrasodium EDTA as a destabilising agent against biofilms should be considered and may aid in the removal of biofilms in the wound ecosystem (Percival et al. 2017) and represents an important component of a programme for patients undergoing biofilm-based wound management (Rhoads et al. 2008).

References

Anon (2004) Ethylenediaminetetraacetic acid (EDTA) and the salts of EDTA: Science Assessment Document for Tolerance Reassessment

Banin E, Brady KM, Greenberg EP (2006) Chelator-induced dispersal and killing of Pseudomonas Aeruginosa cells in a biofilm. Appl Environ Microbiol 72(3):2064–2069

Che Y, Sanderson K, Roddam LF, Kirov SM, Reid DW (2009) Iron-binding compounds impair Pseudomonas Aeruginosa biofilm formation, especially under anaerobic conditions. J Med Microbiol 58(6):765–773

Devine D, Percival R, Wood D, Tuthill T, Kite P, Killington R, Marsh P (2007) Inhibition of biofilms associated with dentures and toothbrushes by tetrasodium EDTA. J Appl Microbiol 103 (6):2516–2524

Donlan RM (2002) Biofilms: microbial life on surfaces. Emerg Infect Dis 8(9):881

Finnegan S, Percival SL (2014) EDTA: an antimicrobial and Antibiofilm agent for use in Wound Care. Advances in Wound Care 4(7):415–421

Francolini I, Donelli G (2010) Prevention and control of biofilm-based medical-device-related infections. FEMS Immunol Med Microbiol 59(3):227–238

Grundler OJ, van der Steen AT, Wilmot J (2005) Overview of the European risk assessment on EDTA in ACS publications

Kimmel CA (1977) Effect of route of administration on the toxicity and teratogenicity of EDTA in the rat. Toxicol Appl Pharmacol 40(2):299–306

Kite P, Hatton D (2001) Antiseptic compositions, methods and systems. US 8541472, EP1628655

Kite P, Eastwood K, Sugden S, Percival SL (2004) Use of in vivo-generated biofilms from hemodialysis catheters to test the efficacy of a novel antimicrobial catheter lock for biofilm eradication in vitro. J Clin Microbiol 42(7):3073–3076

Kite P, Eastwood K, Percival SL. (2005) Assessing the effectiveness of EDTA formulations for use as a novel catheter lock solution for the eradication of biofilms. In Biofilms, persistence and ubiquity, Eds McBain A, Allison D, Pratten J, Spratt D, Upton M and Verran J. Bioline, Cardiff, p 181–190

Koulaouzidou EA, Margelos J, Beltes P, Kortsaris AH (1999) Cytotoxic effects of different concentrations of neutral and alkaline EDTA solutions used as root canal irrigants. J Endod 25(1):21–23

Moreau-Marquis S, O'toole GA, Stanton BA (2009) Tobramycin and FDA-approved iron chelators eliminate Pseudomonas aeruginosa biofilms on cystic fibrosis cells. Am J Respir Cell Mol Biol 41 (3):305–313

Percival SL (2017) Importance of biofilm formation in surgical infection. Br J Surg 104(2):e85–e94

Percival SL, Kite P, Eastwood K, Murga R, Carr J, Arduino MJ, Donlan RM (2005) Tetrasodium EDTA as a novel central venous catheter lock solution against biofilm. Infect Control Hosp Epidemiol 26 (6):515–519

Percival SL, Hill KE, Williams DW, Hooper SJ, Thomas DW, Costerton JW (2012) A review of the scientific evidence for biofilms in wounds. Wound Repair Reg 20(5):647–657

Percival SL, Suleman L, Vuotto C, Donelli G (2015) HCAI, medical devices and biofilms: risk, tolerance and control. J Med Microbiol 64(Pt 4):323–334

Percival SL, Mayer D, Malone M, Swanson T, Gibson D, Schultz G (2017) Surfactants and their role in wound cleansing and biofilm management. J Wound Care (11):26, 680–690

Raad II, Fang X, Keutgen XM, Jiang Y, Sherertz R, Hachem R (2008) The role of chelators in preventing biofilm formation and catheter-related bloodstream infections. Curr Opin Infect Dis 21(4):385–392

Ramage G, Wickes BL, López-Ribot JL (2007) Inhibition on Candida albicans biofilm formation using divalent cation chelators (EDTA). Mycopathologia 164(6):301

Rhoads DD, Wolcott RD, Percival SL (2008) Biofilms in wounds: management strategies. J Wound Care 17 (11):502–508

Schardein J, Sakowski R, Petrere J, Humphrey R (1981) Teratogenesis studies with EDTA and its salts in rats. Toxicol Appl Pharmacol 61(3):423–428

Shanks RM, Donegan NP, Graber ML, Buckingham SE, Zegans ME, Cheung AL, O'Toole GA (2005) Heparin stimulates Staphylococcus aureus biofilm formation. Infect Immun 73(8):4596–4606

Weinberg E (2004) Suppression of bacterial biofilm formation by iron limitation. Med Hypotheses 63 (5):863–865

Wingender J, Neu TR, Flemming H-C (2012) Microbial extracellular polymeric substances: characterization, structure and function. Springer, Berlin

Wolcott R, Costerton J, Raoult D, Cutler S (2013) The polymicrobial nature of biofilm infection. Clin Microbiol Infect 19(2):107–112

Adv Exp Med Biol - Advances in Microbiology, Infectious Diseases and Public Health (2018) 9: 111–126
DOI 10.1007/5584_2018_170
© Springer International Publishing AG 2018
Published online: 17 February 2018

Incidence and Drug Resistance of Zoonotic *Mycobacterium bovis* Infection in Peshawar, Pakistan

Irfan Khattak, Muhammad Hassan Mushtaq, Sultan Ayaz, Sajid Ali, Anwar Sheed, Javed Muhammad, Muhammad Luqman Sohail, Haq Amanullah, Irshad Ahmad, and Sadeeq ur Rahman

Abstract

Prevalence of zoonotic *Mycobacterium bovis* (bTB) disease in human population is underreported from the North of Pakistan. Here, we report on the proportion of human bTB disease among the overall TB patients, drug resistance pattern of bTB isolates, and knowledge, attitude, and practices (KAP)-based analysis of bTB disease. For this purpose, sputum samples from a total of 300 clinically diagnosed TB patients and 100 randomly selected school children suspected of pulmonary TB were processed by culture as well as polymerase chain reaction (PCR) for isolation, identification, and confirmation of *Mycobacterium tuberculosis* (mTB) and bTB species. Isolates of bTB were processed for drug susceptibility tests. Data on KAP regarding TB were obtained on a pretested questionnaire. Sputum-based PCR results indicated that 288/300 (96%) were confirmed as mTB, while 12/300 (4%) were found as bTB diseases. Interestingly, none of the school child was declared positive for either mTB or bTB. Notably, 274/300 (91.3%) positively cultured samples were identified as

I. Khattak
Department of Epidemiology and Public Health, University of Veterinary and Animal Sciences, Lahore, Pakistan

College of Animal Husbandry & Veterinary Sciences, Abdul Wali Khan University, Mardan, Pakistan

M. H. Mushtaq
Department of Epidemiology and Public Health, University of Veterinary and Animal Sciences, Lahore, Pakistan

S. Ayaz and S. ur Rahman (✉)
College of Animal Husbandry & Veterinary Sciences, Abdul Wali Khan University, Mardan, Pakistan
e-mail: Sadeeq@awkum.edu.pk

S. Ali and A. Sheed
Provincial Tuberculosis Reference Laboratory, Peshawar, Khyber Pakhtunkhwa, Pakistan

J. Muhammad
University Diagnostic Lab, University of Veterinary and Animal Sciences, Lahore, Pakistan

M. L. Sohail
University College of Veterinary and Animal Sciences, The Islamia university of Bahawalpur, Bahawalpur, Pakistan

Department of Clinical Medicine and Surgery, University of Veterinary and Animal Sciences, Lahore, Pakistan

H. Amanullah
Department of Clinical Medicine and Surgery, University of Veterinary and Animal Sciences, Lahore, Pakistan

I. Ahmad
School of Biomedical Sciences, University of Leeds, Leeds, UK

mTB, 13/300 (4.3%) as bTB, while 5/300 (1.7%) as mixed containing both. Importantly, except one, all of the bTB isolates were found resistant to pyrazinamide. Surprisingly, most of the bTB isolates (~70%) were found resistant to a broad range of first- and second-line anti-TB drugs. SplitsTree and recombination analysis indicated no evidence of intergenic recombination. Finally, residence, occupation, presence of animals at home, and sleeping alongside animals were found significantly associated with occurrence of bTB disease. To the best of our knowledge, we report for the first time on the high (4%) burden of bTB disease in human TB patients in Peshawar, Pakistan.

Keywords

Drug resistance · Knowledge, attitude, and practices · *Mycobacterium bovis* · Pakistan · Pyrazinamide · Tuberculosis · Zoonosis

1   Introduction

Among the members of the *Mycobacterium tuberculosis* complex (MTBC), *M. bovis* (bTB) is the principal agent of tuberculosis (TB) in domestic and wild animals, whereas *M. tuberculosis* (mTB) is mainly a human pathogen (de Lisle et al. 2002, Grange and Collins 1987, Morris et al. 1994, O'reilly and Daborn 1995). *M. bovis* can be taken in by ingestion, by inhalation, and, less frequently, by contact with mucous membranes and broken skins causing zoonotic TB (bovine tuberculosis) in human. Members—particularly *Mycobacterium tuberculosis* (mTB) and *Mycobacterium bovis* (bTB)—of the MTBC exhibit differences in host preferences and geographical distribution in the development of tuberculosis despite their close genetic relationship (Kasai et al. 2000, Niemann et al. 2000). Of note, TB caused by bTB is radiographically, clinically, and pathologically indistinguishable from TB disease caused by mTB (Grange 2001). Differentiation and accurate diagnosis of the members of the MTBC is essential for epidemiological investigation of human cases and, to some extent, for adequate treatment of the human bTB disease. In 2013, an estimated 9.0 million people developed tuberculosis (TB), and 1.5 million died from the disease. In Pakistan, TB incidence was recorded 0.274%, and prevalence was reported as 0.342% during 2013 (WHO 2014).

The occurrence of bTB disease in humans has been reported low in countries with effective control programs for bovine tuberculosis (bTB) (Hlavsa et al. 2008, Sunder et al. 2009, Rodriguez et al. 2009, Traore et al. 2012, Ojo et al. 2008, Majoor et al. 2011) and remained higher in countries where bTB programs are not being implemented, mainly, due to insufficient resources (Ayele et al. 2004). Although human-to-human transmission of bTB events has been sporadically reported (Sunder et al. 2009), however, considering the increasing incidence of TB globally, epidemiological data on the incidence and impact of bTB on human is crucial. Unfortunately, rigorous control measures, constant vigilance programs, and surveys have been constantly ignored, particularly in the developing countries like Pakistan, largely, due to limited resources (Laniado-Laborin et al. 2014). Available data indicate random reports of widespread presence of bTB disease in animals and human in the south of Pakistan (Javed et al. 2012, Khan and Khan 2007). However, to the best of our knowledge, reports on bTB disease in the Northwestern province, Khyber Pakhtunkhwa, of Pakistan are lacking indicating ignorance of an important issue.

Peshawar, the provincial capital city of Khyber Pakhtunkhwa, Pakistan, remains a gateway to Asia for transport of goods, live animals, and human (see Fig. 1). Peshawar has been central to the migrants from the neighboring country Afghanistan; and those who are living in Peshawar generally maintain businesses and family relations with those in Afghanistan. Trade of meat and even of live animals for food and other purposes remains substantial between Pakistan and Afghanistan. Besides that, substantial expansion of dairy farms in urban and peri-urban area of the city concentrates movement of cattle between herds without such herds being certified free from zoonotic tuberculosis (bTB). A study carried out in 2010 in Peshawar, Pakistan, reported 32.02% (49/153) sputum samples being found positive for

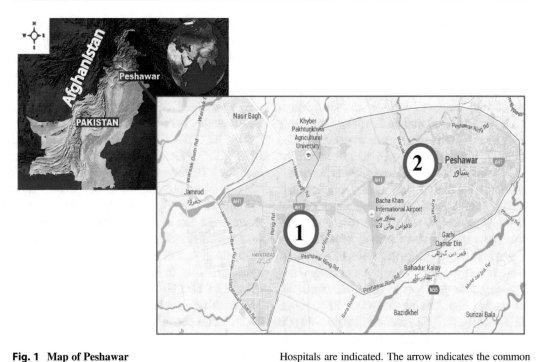

Fig. 1 Map of Peshawar
Peshawar District is encircled, while the two hospitals (1) Hayatabad Medical Complex and (2) Hashtnagri Hospitals are indicated. The arrow indicates the common transport highway used for international transport (Google map was used to draw Fig.1)

TB (Ayaz et al. 2012). Pakistan is ranked sixth among the high-burden TB countries in the world with an estimated 55,000 new TB cases reported every year only in Khyber Pakhtunkhwa (KPK) province (Peshawar the under study city is the capital city of KPK) (http://www.ntp.gov.pk/cmsPage.php?pageID=26). A national TB control program was thus initiated in response and to decline and constrain the incidence of TB in the country. Reference TB labs that were constructed under this initiative, however, are focusing only on the diagnosis of pulmonary TB based on microscopic examination of acid-fast stained sputum samples and radiography (http://www.ntp.gov.pk/cmsPage.php?pageID=26). Consequently, due to lack of guidelines for the differentiation of mTB and bTB, human bTB disease remained underreported from the province. Furthermore, for effective prevention and control of bTB, it is necessary to properly address primary barriers for the prevention of bTB, and hence, it is equally important to understand the perceptions and practices about TB. In this study we report on the incidence of human bTB among samples of active TB cases and KAP about TB disease among these active TB cases. We tested the isolates for drug sensitivity to determine the trend of drug resistance that is emerging around the globe. To the best of our knowledge, we present a comprehensive report for the first time on the burden of bTB disease in human in Peshawar, Pakistan, and identify its links with livestock.

2 Materials and Methods

2.1 Ethics Approval and Consent to Participate

The study and associated protocols were designed based on national ethical legislative rules and approved by the local Ethics Review Committee of UVAS, Lahore, Pakistan, for both animal and human rights. For KAP analysis, all interview participants were first briefed about the aims of the study, and written consent was obtained.

2.2   Study Type Area, Design, and Sampling Strategy

The current study is cross-sectional in nature conducted in Peshawar, Khyber Pakhtunkhwa, carried out between January and September 2015. A total of 300 sputum samples (one sample per patient) from consecutive clinically active adult TB patients (diagnosed at TB centers by routine radiography and microscopic methods) were collected from the city tertiary care hospitals of Peshawar District, the provincial capital of Khyber Pakhtunkhwa, Pakistan (see Fig.1). Of note, these sputum samples were collected prior to medication. A pretested questionnaire was administered to these TB patients to obtain data regarding their demography, socioeconomic status (very high above 150,000 PKR; high 60,000–150,000 PKR; medium 20,000–60,000 PKR; low <10,000–20,000 PKR; very low <10,000 PKR), and knowledge about clinical signs and symptoms of TB, transmission, and social impact of TB (Appendix A) (Khattak et al. 2016).

Since all of the 300 sputum samples were TB adult cases, we wanted to determine incidence of TB in children in Peshawar District. For this purpose, children (aged between 10 and 13 years) with cough (>2 months) in 100 schools were identified, and one child was then selected from each school by draw method. None of these school children were pre-diagnosed nor medicated for TB. All these sputum samples were further analyzed at Provincial Tuberculosis Reference Laboratory, Peshawar, Khyber Pakhtunkhwa, Pakistan, for molecular detection and bacterial isolation.

2.3   Molecular Detection of *M. tuberculosis* and *M. bovis* from Sputum Samples Using Conventional PCR

Each sputum sample was processed for DNA isolation for molecular detection of bTB and mTB species and also cultured for isolation.

Mycobacterial chromosomal DNA was isolated directly from sputum samples using GenoLyse® DNA kit (Hain Lifescience GmbH, Germany) as advised by the manufacturer. Primer pair pncATB-1   ATGCGGGCGTTGATCATCGTC and pncAMT-2 CGGTGTGCCGGAGAAGCGG was used for mTB detection, while primer set JB21   TCGTCCGCTGATGCAAGTGC   and JB22   CGTCCGCTGACCTCAAGAAAG   was used for bTB detection. Detection of mTB was confirmed by observing a PCR amplicon size of 185 bp, while bTB was confirmed by observing an amplicon size of 500 bp. Conditions for PCR were essentially described earlier (Nawaz et al. 2012, Rodriguez et al. 1995). Reagents for PCR were purchased from Thermo Fisher Scientific, USA. The 500 bp amplified PCR products of bTB-positive samples were sequenced by First BASE Laboratories, Malaysia, and the sequences were compared with each other and with already published global sequences using T-Coffee multiple sequence alignment server (di Tommaso et al. 2011). A phylogenetic tree was obtained using mega 6.0 software by employing neighbor-joining method with 1000 bootstrap values.

The split network analysis of the nucleotide sequences of 500 bp PCR product was produced by applying neighbor-net method using SplitsTree4 (Huson and Bryant 2006). The pairwise homoplasy index (*phi*) test (Bruen et al. 2006) integrated into SplitsTree4 (Huson and Bryant 2006) for recombination was performed, and the P value <0.05 indicated recombination event has occurred. Nucleotide diversity and GC contents were determined using MEGA 6 program.

2.4   Culture, Isolation, and Drug Susceptibility of *M. bovis* Clinical Isolates

All 400 sputum samples were processed for isolation of bTB and mTB species. Standard procedures for isolation of bTB and mTB species were followed (Morcillo et al. 2008). Lowenstein-Jensen (LJ) media was used for isolation of mTB, and stone brink (SB) media was used for isolation of bTB. LJ and SB media were purchased from Oxoid, UK. Isolates were subjected to drug susceptibility testing (DST) for at least five first-line anti-TB drugs, i.e., strepto-mycin (STR), isoniazid (INH), rifampicin (RMP), ethambutol (EMB), and pyrazinamide (PZA), and for at least three second-line anti-TB drugs, i.e., amikacin (AMK), capreomycin (CPM), and ofloxacin (OFX). *M. bovis* inoculums for DST were prepared from solid media following manufacturer's instructions, and DST was performed in BACTEC MGIT 960 instrument and interpreted as described earlier (Tortoli et al. 2002). Final drug concentrations used in DST were 1.0 µg/ml for STR, 0.1 µg/ml for INH, 1.0 µg/ml for RMP, 5.0 µg/ml for EMB, 100 µg/ml for PZA, 1 µg/ml for AMK, 2.5 µg/ml for CPM, and 2.0 µg/ml for OFX.

3 Results

3.1 Molecular Detection of bTB and mTB by Targeted PCR Amplification

Our results indicated that 288/300 (96%) of the samples were PCR-positive for mTB, while 12/300 samples (4%) were confirmed positive for bTB (Fig. 2a and b). Interestingly, all sputum samples of 100 school children were revealed PCR-negative for the presence of both mTB and bTB, respectively (results not shown).

3.2 Phylogenetic and Sequence Compositional Analysis

It is known that the 500 bp fragment is quite specific to bTB, while it has never been reported from mTB species. This region is highly conserved among bTB isolates. Multiple sequence alignments of the 12 clinical isolates of the 500 bp PCR product indicated high degree of homology among all isolates; however, there

were slight differences of single bases at random positions (Fig. 2e). Nucleotide composition indicated and average GC contents of 60.8%. Antibacterial (antimycobacterial) drugs are widely used against tuberculosis, and our results indicated that most of the mTB isolates were found resistant to commonly used antimyco-bacterial drugs (see below). Therefore, in order to test whether the current isolates are diverse or clonal in ancestral origin, we tested for recombi-nation. The *phi* test is known as statistically effi-cient and rapid test for analysis of recombination events. The *P* value obtained from *phi* test for all the 12 isolates is 1 suggesting no evidence of significant incidence of recombination across the isolate. Similarly, network analysis also indicated a rectangular-shaped graph suggesting no evi-dence of intergenic recombination between the isolates (Fig. 3).

Phylogenetic analysis of the concatenated sequences of the conserved 500 bp segment of all the 12 isolates indicated that the isolates were clustered into two major groups with further sub-clusters. Altogether, a total of six clusters were observed containing a major cluster with six isolates.

3.3 Isolation and Identification of bTB and mTB by In Vitro Culture

Our results indicated that in consistent with the PCR results, all school children sputum samples did not reveal growth of neither bTB nor mTB species. However, of the adult 300 samples, 274 (91.3%) showed mycobacterial growth on Lowenstein-Jensen (LJ) media indicating mTB isolates, while 13/300 (4.3%) samples showed growth on stone brink (SB) media indicating bTB isolates. More interestingly, 5/300 (1.7%) samples showed growth on both LJ and SB media indicating mixed disease. Finally, 7/300 samples (2.3%) were found contaminated, and 2/300 (0.7%) samples did not show any growth (Fig. 2d and Table-supp. 1).

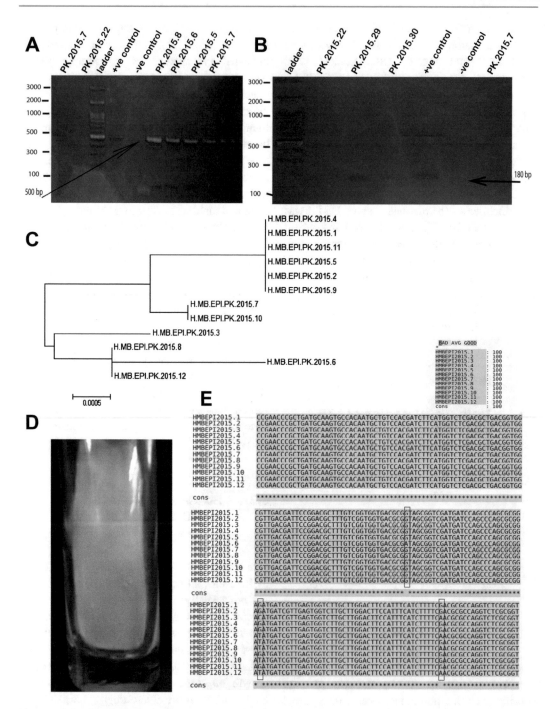

Fig. 2 Molecular detection, sequence, and cultural analysis of mTB and bTB clinical isolates

Above figure indicates (**a**) detection of *M. bovis*-specific 500 bp bands and (**b**) detection of *M. tuberculosis*-specific 185 bp bands. The size of the expected PCR amplicon is calculated with respect to a standard molecular weight ladder that was run in parallel with each sample. (**c**) shows phylogenetic analysis based on the DNA sequence of 500 bp specific band of bTB. (**d**) indicates a representative culture of *M. bovis* clinical isolate. (**e**) indicates sequences alignment of the bTB-specific 500 bp sequenced fragment using T-Coffee alignment. Sequences with variations have been marked with squares. Sequences are not shown to scale, and only those alignments with random variations have been indicated. mTB *Mycobacterium tuberculosis*, bTB *M. bovis*/bovine *tuberculosis*

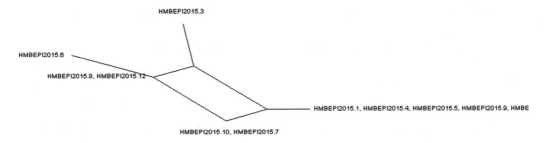

Fig. 3 Split networking
Split network analysis of the concatenated sequences of 500 bp long conserved part specifically found in bTB

species. The network graph indicated a rectangular-shaped structure suggesting no evidence of intergenic recombination

3.4 In Vitro Drug Susceptibility

All these 12 bTB isolates were subjected to drug susceptibility testing (DST). Our results indicated that out of 12 bTB isolates, 11 were resistant to PZA, and only one remained sensitive to PZA. Furthermore, resistance to STR, INH, RMP, OFX, and AMK was shown by 8 (66.7%), 8 (66.7%), 7 (58.3%), 2 (16.7%), and 1 (8.3%) isolates, respectively. Remarkably, none of the samples showed resistance to CPM (Table 2).

3.5 Risk Factors of Zoonotic TB and Demographic and Socioeconomic Conditions of TB Patients

Of the overall 300 human TB patients, Pakistani nationals were 96% (288/300), and Afghan nationals were 4% (12/300) (Table 3). Analysis of demographic and socioeconomic data indicated that associated risk factors of bTB in human were occupation (p 0.005), residence (p 0.001), presence of animal at home (p < 0.0005), and sleeping in animal sheds (p 0.005). Ten out of 12 bTB cases have zoonotic exposure. Importantly, burden of bTB was found higher (6 out of 10) among livestock farmers as compared to other professions (4 out of 10). Similarly, burden of bTB was also found higher among people living in rural areas among those living or sleeping with animals in their sheds (Table 3).

3.6 Knowledge, Attitude, and Practices of TB Patients Regarding TB

Our data indicated that 51.3% considered microorganisms/germs as causative agents of TB disease, while 11.7% stated that TB is hereditary disease. All of these participants had cough for more than 2 weeks along with blood in sputum at some stage. Other self-reported signs and symptoms of TB by the respondents were weight loss (95.3%), night sweating (89.7%), fever (88.3%), chest pain (80.7%), weakness or fatigue (77.0%), no appetite (71.7%), and chills (71.0%) (Table 4). Importantly, bTB was considered as a communicable disease by 82.7% study participants. The perceived mode of bTB transmission mentioned by majority (86%) of the respondents was living with TB patients (Table 5). Of note, a significant number (78.7%) of patients correctly indicated that TB transmission from animal to human can take place through consuming raw meat (78.7%). Similarly, unboiled milk and aerosol routes were also mentioned by 60.3% and 30.3% patients, respectively, as transmission pathway of bTB from animals to humans. Notably, some (31.7%) patients incorrectly stated that contacts with feces, riding, and consumption of cooked meat and boiled milk would also lead to successful transmission of bTB from animals to human (Table 6). Majority of the patients (96.3%) were hopeful that TB is treatable, and the most trusted method of cure was treatment provided by a physician (90.3%) and herbal cures/consulting

Table 1 PCR-based detection, growth, and isolation of mTB and bTB isolates from sputum samples of clinically active TB patients

	PCR results on sputum sample	Growth on LJ result	Growth on SB
bTB	12	0	12
mTB	274	274	0
	2	0	0
	5	5	5
	7	Contaminated	Contaminated

Table 2 Drug susceptibility testing of 12 clinical isolates of bTB

Sample ID	TB disease type	Resistant profile	Susceptible profile
Asm-15	MDR	All first-line drugs, one second line and fluoroquinolone	Two second line
Bur-15	MDR	All first-line drugs	All second line
Gul-15		One first-line drug	Four first line and all second line
Ila-15	MDR	Four first-line drugs	One first line and all second line
Nis-15		One first-line drug	Four first line and all second line
Mum-15		Four first-line drugs and one injectable second-line drug	One first line and two second line
Noo-15	MDR	Four first-line drugs	One first line and three second line
Zia-15		One first-line drug	One first line and three second line
Ras-15	MDR	Four first-line drugs	One first line and three second line
Sai-15	MDR	All first-line drugs, one second-line and fluoroquinolone	Two second line
Sid-15	MDR	All first-line drugs	All three second line
Lal-15		One first-line drug	Four second line and all three second line

hakeem (33.7%). Consulting hakeems are widely practicing in Pakistan mainly through naturally producing wild plants and herbs. Majority (78.3%) of the patients mentioned that TB is preventable, and the most frequently mentioned methods of TB prevention were good nutrition (52.3%) and covering mouth when sneezing or coughing (50.3%) (Table 5).

Our results showed that 59.7% of the patients considered TB as a fatal disease. When we asked that "What feelings they had when were informed as TB-positive?" the responses were fear (29.0%), sadness/hopelessness (28.3%), surprise (26.7%), shame, hurt, and others. When we asked TB patients that "how a person with TB gets on by the community," 17.3% responded that people would not want TB patients to let them play with their children, 14.3% stated that community considers TB patients as unclean, and 12.7%

responded that people keep distance from TB patients.

Our data indicated that majority (94.0%) of the participants always used boiled milk, 5.3% occasionally unboiled milk, while 0.7% regularly used raw milk. Thirty-five participants (11.7%) had TB-infected person at home, and 25 (8.3%) had been in close contact with TB patients. With respect to sources of information about TB, most frequently mentioned sources of information were television (60.3%) and health workers (25.3%).

4 Discussion

Bovine tuberculosis caused by *M. bovis* is increasingly reported from all over the world (WHO 2014), particularly from developing countries (Ayele et al. 2004, WHO 2014, Zaman

Table 3 Bivariate frequency analysis of various parameters in *M. tuberculosis*- and *M. bovis*-infected patients

Characteristics	*M. tuberculosis*-infected patients, N	*M. bovis*-infected patients, n (%)	P value
Nationality			
Pakistani	276	12 (4.2)	0.607
Afghani	12	0 (0.0)	
Age			
<20 years	21	0 (0.0)	0.658
20–30 years	43	3 (6.5)	
31–40 years	62	2 (3.1)	
41–50 years	68	4 (5.6)	
50 years	94	3 (3.1)	
Education level			
Illiterate	34	1 (2.8)	0.713
Middle	85	2 (2.3)	
Secondary	101	5 (4.7)	
Graduate or above	68	4 (5.6)	
Residence			
Urban	213	3 (1.4)	0.001
Rural	75	9 (10.7)	
Socioeconomic status			
Very high	3	0 (0.0)	0.060
High	44	0 (0.0)	
Medium	142	8 (5.3)	
Low	76	1 (1.3)	
Very low	23	3 (11.5)	
Marital status			
Single	37	1 (2.6)	0.902
Married	228	10 (4.2)	
Divorced	6	0 (0.0)	
Widow	17	1 (5.5)	
Occupation			
Livestock farmer	42	6 (12.5)	0.005
Abattoir worker/butcher/veterinarian	2	0 (0.0)	
Others	244	6 (2.4)	
Animals at home			
Cattle/buffalo	67	10 (13.0)	0.000
Others	12	0 (0.0)	
None	209	2 (9.5)	
Sleep in animal shed			
Yes	3	4 (57.1)	0.000
No	285	8 (2.7)	

2010); however, studies focusing on *M. bovis* disease in human and its risk factors in Pakistan are limited. Particularly, human cases of bTB have never been reported from the western province of Khyber Pakhtunkhwa, Pakistan. In the current study, we report on the burden of bTB diseases in human and its associated risk factors related sociodemographic conditions of Khyber Pakhtunkhwa, Pakistan. Furthermore, the drug susceptibility and KAP regarding bTB diseases have been also highlighted. Such kinds of information are useful both for clinicians as well as for policy makers for future control strategies.

In the current study, *M. bovis* DNA was detected in the sputum of 4% adult human TB patients. This reported burden of bTB disease in

Table 4 Self-reported clinical signs and symptoms in TB patients

Signs and symptoms	n (%)
Chronic cough lasts for more than 3 weeks	300 (100.0)
Pain in the chest	242 (80.7)
Coughing up blood or sputum	300 (100.0)
Weakness or fatigue	231 (77.0)
Weight loss	286 (95.3)
No appetite	215 (71.7)
Chills	213 (71.0)
Fever	265 (88.3)
Sweating at night	269 (89.7)

human in Pakistan remained higher as compared to 0.4% in Argentina (de Kantor et al. 2008), 0.8% in Mali (Traore et al. 2012), 0.5–1.5% cases in the United Kingdom (de la Rua-Domenech 2006), 1.2% in Burkina Faso (Sanou et al. 2014), 1–2% in the Unites States (Hlavsa et al. 2008), 1.4% in the Netherlands (Majoor et al. 2011), 2% in France (Sunder et al. 2009), 1.9% in Spain (Rodriguez et al. 2009), 1.6% in Brazil (Silva et al. 2013), and 3% in Ireland (http://www.bovinetb.info/ireland.php). However, interestingly, the current reported burden remained similar to what has been published for other regions of Pakistan (Jabbar et al. 2015). Considering the factor that M. bovis does not predominantly transmit between humans (Sunder et al. 2009), rather the mode of transmission from infected animals to susceptible human is considered efficient (Phillips et al. 2003), it is likely that

Table 5 Participants' knowledge about transmission of TB

	Yes	No	Don't know
	Number (%)	Number (%)	Number (%)
Is TB a communicable disease?	248 (82.7)	36 (12.0)	16.0 (5.3)
TB transmitted through			
Cough or sneeze	162 (54.0)	90 (30.0)	48 (16.0)
Handshake	35 (11.7)	211 (70.3)	54 (18.0)
Sexual relations	13 (4.3)	184 (61.3)	103 (34.3)
Sharing food	95 (31.7)	153 (51.0)	52 (17.3)
Living with patient Tb	258 (86.0)	7 (2.3)	35 (11.7)
Excess work	57 (19.0)	215 (71.7)	28 (9.3)
TB transmitted from animals to human through			
Consumption of unboiled milk	181 (60.3)	112 (37.3)	7 (2.3)
Consumption of raw meat	236 (78.7)	59 (19.7)	5 (1.7)
Through aerosol	91 (30.3)	206 (68.7)	3 (1.0)
Others (boiled milk, cooked meat, riding, contact with feces)	95 (31.7)		
Can TB be prevented?	235 (78.3)	41 (13.7)	24 (8.0)
Preventive methods of TB			
Good nutrition	157 (52.3)	90 (30.0)	53 (17.7)
Closing windows	45 (15.0)	82 (27.3)	173 (57.7)
Covering your mouth when sneezing or coughing	151 (50.3)	94 (31.3)	55 (18.3)
Not sharing food	89 (29.7)	169 (56.3)	42 (14.0)
Avoiding use of utensils used by patient TB	74 (24.7)	163 (54.3)	63 (21.0)
Is TB treatable?	289 (96.3)	3 (1.0)	8 (2.7)
Treatment method of TB			
Adhering to treatment provided by physician	271 (90.3)	16 (5.3)	13 (4.3)
Herbal cures/consulting hakeem	101 (33.7)	176 (58.7)	23 (7.7)
Resting but not taking medications	53 (17.7)	233 (77.7)	14 (4.7)
Prayers/eating well/others	96 (32.0)	190 (63.3)	14 (4.7)

Table 6 Participants' knowledge and habit about consumption of milk

Characteristics	*M. tuberculosis*-infected patients, N	*M. bovis*-infected patients, n (%)	P value
Milk consumption habits			
Always use boiled milk	273	9 (3.19%)	0.001
Always use raw milk	1	1 (50%)	
Occasionally use raw milk	15	1 (6.66%)	

the identified burden of the bTB disease indicates zoonotic transmission. Nevertheless, this is important to know that all these isolates were actually transmitted directly from animals to human and not as a result of clonal expansion of human-adapted *M. bovis* species that had once been transmitted from cattle. We are currently working on further characterization of these strains and related risk factors to further explain this possibility. Furthermore, the current study has limitations such as we have not investigated the intrinsic factors of bTB responsible for dissemination or incidence of TB infection. Moreover, unpasteurized milk and properly uncooked meat may also be high-risk factors responsible for TB diseases that have not been investigated. In conclusion our findings indicate a higher burden of circulating strain of *M. bovis*.

Livestock farmers and those involved in rearing cattle and buffaloes at home were found more infected by bTB as compared to those of other professions and with no physical contacts with domestic animals. Similar findings have been reported in other parts of the world such as in Mexico, 11.8% of the sputum samples from cattle farm workers were found positive for *M. bovis* (Milian-Suazo et al. 2010) and 65% of *M. bovis*-infected patients in Argentina were occupationally exposed (Cordova et al. 2012). In Bangladesh "rearing of livestock in household" was identified as risk factor of zoonotic TB (Rahman et al. 2015). In our study, rural residence was found significantly associated with occurrence of zoonotic TB. Cordova et al. (2012) have reported that 31% of the zoonotic TB patients in Buenos Aires, Argentina, were living in rural areas. Rural residents more often keep livestock and more often shared dwelling with animals than urban (Kilale et al. 2015).

In the current study, we found that 66.7% (8/12) of *M. bovis* isolates were resistant to isoniazid and rifampicin. However, in contrast, resistance pattern of *M. bovis* to isoniazid and rifampicin has been reported in the range of 9.5–16.1%, respectively, in other parts of the world (Bobadilla-Del VALLE et al. 2015). Similarly, a lower number of isolates (28.5%) were found resistant to isoniazid and rifampicin among the clinical isolates in Ireland (Mclaughlin et al. 2012). Furthermore, in Mexico, three of the six *M. bovis* strains were found to be MDR with no genetic relatedness among them. In our study, the observed high level of resistance to antimycobacterial drugs correlates with our findings of high burden of circulating strains of bTB among humans.

About half (51.3%) of the participants stated that microorganisms/bacteria are the cause of TB. These numbers of people are comparatively high as compared to other studies conducted in different parts of the world exemplified by 22.9% in Southwestern Ethiopia and 42.8% in Bangladesh (Tolossa et al. 2014, Rana et al. 2015). Public awareness regarding the interspecies transmission of TB-causing agents between human and animals held importance for control and prevention. In this study, it was encouraging that 82.7% study participants were aware of this scenario. To prevent bovine tuberculosis and its cross dissemination to human, a successful control and eradication scheme is essential. However, eradication campaigns usually have been hampered by a number of factors such as low socioeconomic standards, low and no availability of health facilities, poor health policies, and lack of rapid screening, identification, treatment/eradication, and containment facilities. Overall, the current study reports on the high burden of bTB disease among human TB patients in Peshawar District, Pakistan, and identifies low level of knowledge and understanding of this important zoonotic disease.

5 Conclusion

In the current study, we showed evidences of *M. bovis* disease in human TB patients in Peshawar District, Pakistan. Keeping in view the abrupt increase in livestock farming in the area, lack of effective screening and certification program for bTB in livestock and animal farms, and absence of active surveillance and vigilance program in human and animals, it is highly recommended that a comprehensive and systemic surveillance program of bovine Tb in human as well as in animals should be initiated in the province.

Author's Contribution I. K., S. A, A.S, and A. H performed sampling and processed them for isolation. M.H.M., SA., M.L.S., M.H.M.,

and SUR conceived, designed, and supervised the study and drafted the manuscript with IK, and M.I. J. M., A. H., H.A., S.U.R., and IK performed PCR and analyzed the results.

Funding and Acknowledgments Financial support to this research was provided by the Higher Education Commission, Pakistan, under Indigenous 5,000 PhD Fellowship Program, Grant Number 112–23265-2AV1–073. The authors acknowledge the generous technical support of medical administration and laboratory staff of Hayatabad Medical Complex, Peshawar; Hashtnagri General Hospital; staff of the Provincial TB Reference Laboratory, Hayatabad Medical Complex, Peshawar, Pakistan; and Department of Animal Health, the University of Agriculture, Peshawar, Pakistan.

Above table indicates the successful number of clinical isolates that were cultured on mycobacterial specific media (LJ and SB). Of the 300 sputum samples, only 2 samples could not successfully grow, while 7 were contaminated and were difficult to declare positive or negative.

Appendix A

Questionnaire used for Human Tuberculosis Patients

M. bovis
M. tuberculosis

Name		CNIC or cell#			
Address				Urban	Rural
Age in years	<20	21–30	31–40	41–50	Above 50
Education	Illiterate	Middle	Secondary	Graduate	
Socioeconomic status	V. high	High	Moderate	Low	V. low
Marital status	Single	Married	Divorced	Widower	

1. **Nationality:** Pakistani   Afghani   Others (specify)

2. **Milk consumption habits:**

Always use boiled milk	
Always use raw milk	
Routinely use boiled milk and occasionally raw milk	

3. Occupation
   (a) Livestock farmer   (b) Abattoir worker/butcher/veterinarian   (c) Others (specify)

_____

4. Animals at home?
   If yes: (a) Cattle/buffalo   (b) Others (specify) _____   (c) none

5. Do you sleep in animal shed?   Yes   No

6. Have you ever contracted tuberculosis before?   Yes   No

7. Is there any tuberculosis-infected person at your home?   Yes   No

8. Have you ever had close contact with anyone who was sick with tuberculosis?   Yes   No

9. Is TB a communicable disease?   Yes 248   No   Don't know 16

10. Cause of TB?   (a) Microorganisms   (b) Hereditary   (c) Don't know

11. Health status of the patient

SNO	Signs and symptoms	Yes	No
1	Chronic cough lasts for more than 3 weeks		
2	Pain in the chest		
3	Coughing up blood or sputum		
4	Weakness or fatigue		
5	Weight loss		
6	No appetite		
7	Chills		
8	Fever		
9	Sweating at night		

12. Tuberculosis is contracted by:

	Yes	No	Don't know
Cough or sneeze			
Handshake			
Sexual relations			
Sharing food			
Living with patient Tb			
Excess work			

13. Tuberculosis can be transmitted from animals to human through:

	Yes	No	Don't know
Consumption of unboiled milk			
Consumption of raw meat			
Through aerosol			
Others (boiled milk, cooked meat, riding, contact with feces)			

14. Is TB preventable?   (a) Yes   (b) No   (c) Don't Know

15. Tuberculosis can be prevented by:

	Yes	No	D/K
Good nutrition			
Closing windows			
Covering your mouth when sneezing or coughing			
Not sharing food			
Avoiding use of utensils used by patient TB			

16. IS TB treatable?   **Yes**   **No**   **Don't know**

17. People affected by TB can be cured by:

	Yes	No	Don't know
Adhering to treatment provided by physician			
Herbal cures/consulting hakeem			
Resting but not taking medications			
Prayers/eating well/others			

18. What feelings do you have when presented with a diagnosis of TB?

(a) Fear

(b) Surprise

(c) Shame

(d) Sadness/hopelessness

(e) Others (hurt, do not fear, etc.)

19. With whom a diagnosis of TB would be shared (can choose multiple answer)?

(a) Physician

(b) Partner

(c) Parent

(d) Offspring

(e) Friends or relation

(f) No person

20. How serious a disease is TB?

(a) Very serious

(b) Somewhat serious

(c) Not very serious

(d) Don't know

21. From what sources did you get information about tuberculosis? (can choose multiple answer)

(a) Newspapers

(b) Radio

(c) TV

(d) Internet

(e) Pamphlets and posters

(f) Health worker

(g) Family member, friend, or neighbor

(h) Others (specify)

22. Community attitudes toward patient tuberculosis (can choose multiple answer)

(a) People keep distance from those with TB

(b) People do not want those with TB playing with their children

(c) People do not want to talk to others with TB

(d) People try not to touch others with TB

(e) People think that those with TB are unclean

(f) Others (specify)

23. Do you know that TB treatment is free?

(a) Yes   (b) No

References

Ayaz S, Nosheen T, Khan S, Khan SN, Rubab L, Akhtar M (2012) Pulmonary tuberculosis: still prevalent in human in Peshawar, Khyber Pakhtunkhwa, Pakistan. Tuberculosis (TB) 10:39–41

Ayele WY, Neill SD, Zinsstag J, Weiss MG, Pavlik I (2004) Bovine tuberculosis: an old disease but a new threat to Africa. Int J Tuberc Lung Dis 8:924–937

Bobadilla-Del VALLE M, Torres-Gonzalez P, Cervera-Hernandez ME, Martinez-Gamboa A, Crabtree-Ramirez B, Chavez-Mazari B, Ortiz-Conchi N, Rodriguez-Cruz L, Cervantes-Sanchez A, Gudino-Enriquez T, Cinta-Severo C, Sifuentes-Osornio J, Ponce de Leon A (2015) Trends of Mycobacterium bovis isolation and first-line anti-tuberculosis drug susceptibility profile: a fifteen-year laboratory-based surveillance. PLoS Negl Trop Dis 9:e0004124

Bruen TC, Philippe H, Bryant D (2006) A simple and robust statistical test for detecting the presence of recombination. Genetics 172:2665–2681

Cordova E, Gonzalo X, Boschi A, Lossa M, Robles M, Poggi S, Ambroggi M (2012) Human Mycobacterium bovis infection in Buenos Aires: epidemiology, microbiology and clinical presentation. Int J Tuberc Lung Dis 16:415–417

de Kantor IN, Ambroggi M, Poggi S, Morcillo N, Telles MADS, Ribeiro MO, Torres MCG, Polo CL, Ribón W, García V (2008) Human Mycobacterium bovis infection in ten Latin American countries. Tuberculosis 88:358–365

de la Rua-Domenech R (2006) Human Mycobacterium bovis infection in the United Kingdom: incidence, risks, control measures and review of the zoonotic aspects of bovine tuberculosis. Tuberculosis 86:77–109

de Lisle G, Bengis R, Schmitt S, O'brien D (2002) Tuberculosis in free-ranging wildlife: detection, diagnosis and management. Revue scientifique et technique (International Office of Epizootics) 21:317–334

di Tommaso P, Moretti S, Xenarios I, Orobitg M, Montanyola A, Chang JM, Taly JF, Notredame C (2011) T-Coffee: a web server for the multiple sequence alignment of protein and RNA sequences using structural information and homology extension. Nucleic Acids Res 39:W13–W17

Grange J, Collins C (1987) Bovine tubercle bacilli and disease in animals and man. Epidemiol Infect 99:221–234

Grange JM (2001) Mycobacterium bovis infection in human beings. Tuberculosis (Edinb) 81:71–77

Hlavsa MC, Moonan PK, Cowan LS, Navin TR, Kammerer JS, Morlock GP, Crawford JT, Lobue PA (2008) Human tuberculosis due to Mycobacterium bovis in the United States, 1995-2005. Clin Infect Dis 47:168–175

Huson DH, Bryant D (2006) Application of phylogenetic networks in evolutionary studies. Mol Biol Evol 23:254–267

Jabbar A, Khan J, Ullah A, Rehman H, Ali I (2015) Detection of Mycobacterium tuberculosis and Mycobacterium bovis from human sputum samples through multiplex PCR. Pak J Pharm Sci 28:1275–1280

Javed MT, Ahmad L, Feliziani F, Pasquali P, Akhtar M, Usman M, Irfan M, Severi G, Cagiola M (2012) Analysis of some of the epidemiological risk factors affecting the prevalence of tuberculosis in buffalo at seven livestock farms in Punjab Pakistan. Asian Biomed 6:35–42

Kasai H, Ezaki T, Harayama S (2000) Differentiation of phylogenetically related slowly growing mycobacteria by their gyrB sequences. J Clin Microbiol 38:301–308

Khan IA, Khan A (2007) Prevalence and risk factors of bovine tuberculosis in Nili Ravi buffaloes in the Punjab, Pakistan. Ital J Anim Sci 6:817–820

Khattak I, Mushtaq MH, Ahmad MuD, Khan MS, Chaudhry M, Sadique U (2016) Risk factors associated with Mycobacterium bovis skin positivity in cattle and buffalo in Peshawar, Pakistan. Trop Anim Health Prod 48:479–485

Kilale AM, Ngadaya E, Kagaruki GB, Lema YL, Muhumuza J, Ngowi BJ, Mfinanga SG, Hinderaker SG (2015) Experienced and perceived risks of mycobacterial diseases: a cross sectional study among agropastoral communities in northern Tanzania. PLoS One 10:e0130180

Laniado-Laborin R, Muniz-Salazar R, Garcia-Ortiz RA, Vargas-Ojeda AC, Villa-Rosas C, Oceguera-Palao L (2014) Molecular characterization of Mycobacterium bovis isolates from patients with tuberculosis in Baja California, Mexico. Infect Genet Evol 27:1–5

Majoor CJ, Magis-Escurra C, van Ingen J, Boeree MJ, van Soolingen D (2011) Epidemiology of Mycobacterium bovis disease in humans, The Netherlands, 1993-2007. Emerg Infect Dis 17:457–463

Mclaughlin AM, Gibbons N, Fitzgibbon M, Power JT, Foley SC, Hayes JP, Rogers T, Keane J (2012) Primary isoniazid resistance in Mycobacterium bovis disease: a prospect of concern. Am J Respir Crit Care Med 186:110–111

Milian-Suazo F, Perez-Guerrero L, Arriaga-Diaz C, Escartin-Chavez M (2010) Molecular epidemiology of human cases of tuberculosis by Mycobacterium bovis in Mexico. Prev Vet Med 97:37–44

Morcillo N, Imperiale B, Palomino JC (2008) New simple decontamination method improves microscopic detection and culture of mycobacteria in clinical practice. Infect drug resist 1:21–26

Morris R, Pfeiffer D, Jackson R (1994) The epidemiology of Mycobacterium bovis infections. Vet Microbiol 40:153–177

Nawaz A, Chaudhry ZI, Shahid M, Gul S, Khan FA, Hussain M (2012) Detection of Mycobacterium tuberculosis and Mycobacterium bovis in sputum and blood samples of human. J Anim Plant Sci 22 (2 suppl):117–120

Niemann S, Harmsen D, Rusch-Gerdes S, Richter E (2000) Differentiation of clinical Mycobacterium tuberculosis complex isolates by gyrB DNA sequence polymorphism analysis. J Clin Microbiol 38:3231–3234

O'reilly LM, Daborn C (1995) The epidemiology of Mycobacterium bovis infections in animals and man: a review. Tuber Lung Dis 76:1–46

Ojo O, Sheehan S, Corcoran GD, Okker M, Gover K, Nikolayevsky V, Brown T, Dale J, Gordon SV, Drobniewski F, Prentice MB (2008) Mycobacterium bovis strains causing smear-positive human tuberculosis, Southwest Ireland. Emerg Infect Dis 14:1931–1934

Phillips C, Foster C, Morris P, Teverson R (2003) The transmission of Mycobacterium bovis infection to cattle. Res Vet Sci 74:1–15

Rahman MM, Noor M, Islam KM, Uddin MB, Hossain FMA, Zinnah MA, Mamun MA, Islam MR, EO SK, Ashour HM (2015) Molecular diagnosis of bovine tuberculosis in bovine and human samples: implications for zoonosis. Future Microbiol 10:527–535

Rana M, Sayem A, Karim R, Islam N, Islam R, Zaman TK, Hossain G (2015) Assessment of knowledge regarding tuberculosis among non-medical university students in Bangladesh: a cross-sectional study. BMC Public Health 15:716

Rodriguez E, Sanchez LP, Perez S, Herrera L, Jimenez MS, Samper S, Iglesias MJ (2009) Human tuberculosis due to Mycobacterium bovis and M. caprae in Spain, 2004-2007. Int J Tuberc Lung Dis 13:1536–1541

Rodriguez JG, Mejia GA, del Portillo P, Patarroyo ME, Murillo LA (1995) Species-specific identification of Mycobacterium bovis by PCR. Microbiology 141:2131–2138

Sanou A, Tarnagda Z, Kanyala E, Zingué D, Nouctara M, Ganamé Z, Combary A, Hien H, Dembele M, Kabore A, Meda N, van de Perre P, Neveu D, Bañuls AL, Godreuil S (2014) Mycobacterium bovis in Burkina Faso: epidemiologic and genetic links between human and cattle isolates. PLoS Negl Trop Dis 8:e3142

Silva MR, Rocha ADA S, da Costa RR, de Alencar AP, de Oliveira VM, Fonseca Junior AA, Sales ML, Issa MDE A, Filho PM, Pereira OT, Dos Santos EC, Mendes RS, Ferreira AM, Mota PM, Suffys PN, Guimaraes MD (2013) Tuberculosis patients co-infected with Mycobacterium bovis and Mycobacterium tuberculosis in an urban area of Brazil. Mem Inst Oswaldo Cruz 108:321–327

Sunder S, Lanotte P, Godreuil S, Martin C, Boschiroli ML, Besnier JM (2009) Human-to-human transmission of tuberculosis caused by Mycobacterium bovis in immunocompetent patients. J Clin Microbiol 47:1249–1251

Tolossa D, Medhin G, Legesse M (2014) Community knowledge, attitude, and practices towards tuberculosis in Shinile town, Somali regional state, eastern Ethiopia: a cross-sectional study. BMC Public Health 14:804

Tortoli E, Benedetti M, Fontanelli A, Simonetti MT (2002) Evaluation of automated BACTEC MGIT 960 system for testing susceptibility of Mycobacterium tuberculosis to four major Antituberculous drugs: comparison with the radiometric BACTEC 460TB method and the agar plate method of proportion. J Clin Microbiol 40:607–610

Traore B, Diarra B, Dembele BP, Somboro AM, Hammond AS, Siddiqui S, Maiga M, Kone B, Sarro YS, Washington J, Parta M, Coulibaly N, M'baye O, Diallo S, Koita O, Tounkara A, Polis MA (2012) Molecular strain typing of Mycobacterium tuberculosis complex in Bamako, Mali. Int J Tuberc Lung Dis 16:911–916

WHO (2014) Global tuberculosis report. World Health Organization, Geneva

Zaman K (2010) Tuberculosis: a Global Health problem. J Health Popul Nutr 28:111–113

Adv Exp Med Biol - Advances in Microbiology, Infectious Diseases and Public Health (2018) 9: 127–130
DOI 10.1007/978-3-319-79017-6
© Springer International Publishing AG, part of Springer Nature 2018

Index

Printed in the United States
By Bookmasters